An Observer's Guide to the

Geology of Prince William Sound, Alaska

By Jim Lethcoe
Preface by Steve W. Nelson
Illustrations by the author

Prince William Sound Books: Valdez, Alaska.

Cover photo: Turbidite sequences and Surprise Glacier, Harriman Fiord, Marti Miller USGS.

ISBN 1-877900-001

Table of Contents

Acknowledgements
Author's Foreword
Preface
Maps
Timeline

Chapter 1. **The Earth in Motion** — 1
    Continental Drift — 3
    Plate Tectonics — 6
    The Restless Plates — 14
    The Aleutian Trench and Southern Alaska — 20
    Terranes of southern Alaska — 32

Chapter 2. **Geological History of Prince William Sound
    and Southern Alaska** — 46
    Carboniferous and Permian Beginnings — 46
    The Birth and Development of Wrangellia — 47
    The Peninsular Terrane &Formation of the Superterrane — 50
    Cretaceous Collision and the Birth of the Chugach/Prince
        William Sound Terrane — 56
    The Paleocene Development of the Orca Group — 60
    Paleocene/Eocene Northward Movement of the Superterrane — 66
    The Eocene Flight Northward — 67
    The Final Collision — 69
    Oligocene Plutonism — 73
    Miocene Uplift — 74
    Continued Pliocene Uplift — 76
    Pleistocene Glaciation — 77
    Holocene Retreat — 79
    Footnotes — 82

Chapter 3. **Observing Prince William Sound's Rocks** — 86
    A. Introduction — 86
    B. Observing Prince William Sound's Volcanic Rocks — 87
        Pillow Basalts — 87
        Sheeted Dikes — 89
        Massive Flows — 90
        Ophiolites — 91
    C. Observing Plutonic Rocks — 92

D. Observing Metamorphic Rocks	97
E. Observing Sedimentary Rocks	102
Turbidites	102
The Bouma Sequences	104
Conglomerates	107
Rip-Up	110
Sole Markings	111
Flute Casts	112
Tool Marks	112
Load Casts	115
Concretions	115
Fossils	117
Bedding	120
Folding and Faulting	121
Folds	122
Faults	123
F. Observing Ore inerals in Prince William Sound	127

Chapter 4. **Field Guide to the Rocks of Prince William Sound** 131

I. Passage Canal Area 131

Whittier	132	Lansing Mine	134
Billings Glacier	132	Ziegler Cove	135
Poe Valley	133	Point cochrane	136
Port Wells	134	South arm of Surprise Cove	136

II. Port Wells and Harriman Fiord

Mineral King Mine	137	Serpentine Cove	140
Granite Mine	138	Coxe Glacier	141
Hobo Bay to Harrison Lagoon	139	Coghill Point	141
Point Doran Peninsula	140	Golden	142
Harriman Glacier	140		

III. Northern Islands 142

Esther Island	142	Culross Island	145
Fisherman's Anchorage	144	Culross Bay	145
East side of Esther Passage	144	Culross Pluton	146
Grommet Cove	144	South Culross Passage	147

Perry Island Pluton	147
Lone Island	147

IV, . Northern Sound   149

Squaw Bay	149	Naked, Peak and Storey I.	153
Contact Fault	149	Bass Harbor	153
Cascade Bay	150	Glacier Island	154
Cow Pens Anchorage	150	Long Bay	155
Miner's Bay	151	Granite Cove	155
Unakwik Mineralized Zone	152	Great Nunatak	156
Cedar Bay Pluton	153		
Outpost Island	153		

V. Valdez Area   157

Old Valdez	158	Valdez Arm Conglomerates	168
Ramsay-Rutherford Mine	160	Johnston Bay	168
Mineral Creek	160	Galena Bay	169
Cliff Mine	162	Galena Bay Mining Co.	173
Midas Mine	163	Ellamar Mine	173
Jack Bay Fault	166	General Scenario	174
Sawmill Bay	168		

VI. Cordova Area   176

Tatitlek Narrows	176	Fidalgo Mining Co. Mine	181
Bligh Island	177	Irish Cove	181
Boulder Bay	177	Porcupine Point	181
Bidarka Point	177	Sheep Bay Pluton	181
Landlocked Bay	177	Comfort Cove	182
Threeman Mine	179	Simpson Bay	183
Landlocked Bay Mining Co.	179	Bomp Point	183
Hemple Prospect	179	Cordova	183
Standard Copper Mins Co.	180	Evak Ridge	184
Landlocked Bay Fault	180	Point Whitshed	184
Fish Bay	180	Canoe Passage	184
Sunny Cove	181	Cedar Bay	184
Fidalgo Marble	181		

VII. Montague Belt — 185

Double Bay — 187
Johnstone Point — 187
Shelter Bay — 187
Deer Cove — 187
Port Etches — 187
Garden Cove — 187
Zaikov Harbor and Rocky Bay — 187

Green Island — 188
Gibbon Anchorage — 189
Stockdale Harbor — 189
Point Gilmore — 189
Hanning Bay — 190
MacLeod Harbor — 190
Patton Bay Fault — 190

VIII. Knight Island Group — 193

Point Eleanor — 195
Lower Passage — 195
Drier Bay — 195
Port Audrey — 196
Squire Island Anchorage — 196

Little to Mummy Bay — 196
Discovery Point — 196
The Pandora Group — 197
Rua Cove Mine — 197
Bay of Isles — 197

IX. Southwest Islands — 199

Sleepy Bay — 199
Latouche Island — 200
Wilson Bay — 200
Latouche — 200
Powder Point — 202
The Beatson Fault — 202
Elrington Island — 202

Fox Farm Anchorage — 202
Evans Island — 203
Shelter Bay — 203
Bainbridge Fault — 203
Hogg Bay Conglomerates — 203
Minke Cove — 203
Orca Cove — 203

X. Western Sound — 204

Kake Cove — 205
Dangerous Passage — 205
Chenega Island — 205
Eshamy Pluton — 206
Granite Bay — 206
Point Nowell — 206

Main Bay — 207
Port Nellie Juan — 207
Nellie Juan Pluton — 207
Deep Water Bay — 207
Greystone Bay — 207
Kings Bay — 207

Acknowledgements:

I would like to thank Nancy Simmerman, Glenn Hershberger, Greg Mallory, Dorothy Clifton and Julie Dumoulin, Marti Miller and Steve Nelson and the USGS for the use of their photographs.

I am especially grateful to Cambridge University Press, John Wiley and Sons, W.H. Freeman and Springer-Verlag for permission to use graphics that have appeared in their publications.

The advice of general readers Mike Rental, Chris Brann, Tim Jones, Andy Embeck, Nancy Lethcoe and Peter and Barbara Sokolov have been especially helpful in preparing this manuscript. I would also like to thank Dave Janka for showing us around Glacier Island.

Special thanks go to geologists Travis Hudson, Dwight Bradley and Steve Nelson for reviewing parts of this manuscript. Any errors are, of course, the author's. Steve Nelson and Joe Kurtak have been especially helpful in aiding in the identification of rocks and fossils collected in the Sound.

Once again, it is a pleasure to extend our thanks to the Valdez City Council for its support of the Valdez Museum and Library and our appreciation to the library and museum staffs for their considerable help.

I am especially grateful to Steve Nelson of the USGS for his kindness and patience in answering my many questions regarding the geology of Prince William Sound.

Author's Foreword

Two years ago, I innocently rapped on the office door of USGS geologist, Steve Nelson, and naively informed him that I intended to write a book on the geology of Prince William Sound and wondered if I could ask him a few questions. Steve, who had led the RARE II project mapping the geology of the Sound, is acknowledged to be the local expert. There was only one slight problem: although I had always been interested in geology and had done a bit of reading in the subject, I had no formal training.

I asked several questions and soon Steve was launched into a lively discussion of the Sound's geology in terms of modern plate tectonics. We pored over some geological maps; he drew numerous diagrams and sketches and showed me a chart of fossil evidence he had compiled; but mostly he just talked. And he talked as only a man consumed by the love of his subject can talk. Assuming that I knew more than I really knew, he spoke in terms familiar to himself and his colleagues. His animated explanations issued forth in a dizzying swirl of a new and strange vocabulary: Maestrichtian turbidites were accreting in an ancient subduction trench, black smokers were spewing forth sulfide-rich precipitates near an oceanic rift, terranes and ophiolites were obducting onto the continental margin. Yet, through the fog of technical terms, I was able to catch occasional glimpses of awesome expanses of time, of crustal fragments transported over great distances, of the ponderous but violent forces that had formed the rocks of Prince William Sound. Two hours later, leaving Steve's office in a bit of a daze, I resolved to do some more reading, learn the language, and return next time as a native speaker.

The reader of this book must make a similar commitment. Modern geology is a highly technical field with its own specialized vocabulary that must be mastered to enjoy the not insignificant reward of being able to comprehend one's world just a little bit better. I have attempted to ease the reader's journey somewhat by employing evocative rather than literal language, by defining technical terms in the text, italicizing essential vocabulary, repeating key terms and concepts throughout, employing numerous photographs and graphics, and including a glossary of technical terms at the end.

The geological time-scale with its specialized classification of eras, epochs, and periods can create especial problems for the uninitiated. I have therefore, often endeavored to spell out these periods in numbers of years followed by the abbreviation *Mya* (million years ago). For quick reference, I have included a time-line summary of this scheme at the front of this book along with a graphic summary of hypothesized plate movements for the past 150 million years. To give these otherwise abstract terms some content and to help orient the reader in time, I have included short vignettes of each period in the geological history section tracing certain familiar themes such as the birth and death of the dinosaurs, and the evolution of plants, animals, mammals and man.

The lay reader should be aware that the Geological History section is only a "scenario" of how things might have happened. The evidence at present concerning the complex geological history of Southern Alaska and Prince William Sound is so incomplete (and often internally inconsistent) that its writing must be left to future geologists. However, we are living in an exciting time as a somewhat hazy picture is now just beginning to emerge. It is as if we have been presented a jig-saw puzzle with half the pieces missing. Different geologists arrange the pieces in slightly different ways depending on the assumptions they make regarding how the the various pieces fit together. In each case, a picture emerges roughly similar to the others in its general outline but often differing dramatically in its details.

It is hoped that by the time the general reader reaches the field guide section of this book that he will have acquired the necessary concepts and vocabulary not only to observe and identify rocks and formations in Prince William Sound but more importantly will have learned to *understand* them. It is hoped that walking a beach, he will be able to see not just rocks but records of past events. For instance, behind a lowly piece of sandstone lying on the beach, there may loom the story of an ancient volcanic arc erupting 200 million years ago near the ancient equator. Its lava was then worn by time's countless eons into a fine sediment to be subsequently swept by bottom hugging, turbidity currents across the floor of some ancient subduction trench. Slowly, these sediments were carried down into the trench where they were buried and transformed into rock. These buried rocks were then transported 1300 miles along the continental margin aboard the Kula Plate where they were uplifted earthquake by earthquake until they occupied a lofty summit in the Chugach Mountains. Freeze-thaw action plucked the rock from the summit dumping it onto a glacier which transported it to its present position here on this beach. This is more than just a rock, this is a rock with a history and a meaning!

Preface

An *Observer's Guide to the Geology of Prince William Sound, Alaska* consists of four parts: part one deals with some basic geological concepts helpful to the reader unfamiliar with geology, parts two and three cover the geologic history and rocks in the area, and part four outlines a field guide to 124 specific localities where geologic features can be observed.

Basic to the understanding of the geology of Prince William Sound, or for that matter understanding the geology of most of the Earth, is the theory of plate tectonics. Using this theory, geologists recognize that the Earth's crust consists of some ten or twelve plates bounded by major fractures or faults, much like an egg shell that has been fractured. The main difference from the egg analogy is that the Earth's plates move. The margins of plates are destroyed along some of the fractures, material is added along others, and as the plates slip by their neighbors, the rocks are bent and broken. What all this means is that rocks formed in one area of the Earth in the past may now be found in another area.

This guide presents the author's interpretation of the geologic migration of the rocks found in Prince William Sound. To do this he needed to integrate concepts from several geologic disciplines. Paleontology, the study of past life forms, provided limits on the age of rocks and conditions of past environments; geophysics, especially paleomagnetism, provided clues to the nature and migrations of segments of the Earth's crust; structural geology helped unravel the complex deformations that affected the rocks of the moving plates; sedimentology suggested processes that took place during deposition of the sedimentary rocks; and igneous petrology provided insight into the development of the granite plutons and submarine volcanic rocks that are important to the understanding of the mineral deposits historically mined in the area. Explorers using this guide will be able to see such things as a 66 million year old fossil clam, ruins of gold and copper mines, folded rock layers, and the intrusion of once molten (800°C) granite.

This guide is an outgrowth of the author's strong interest in the natural history of Prince William Sound after living and guiding in the area for 20 years. Although not a geologist, the author has pursued the study of geology of the area and produced a creditable work in just two years of research. It has been my pleasure to help guide him through the various and sometimes conflicting geologic views that have been published, and to watch him gain an appreciation of the science of geology and the rocks of Prince William Sound. I think that the reader will share this appreciation.

Steven W. Nelson, Geologist
Anchorage, 1990

Plate Movements 150 Mya to the Present

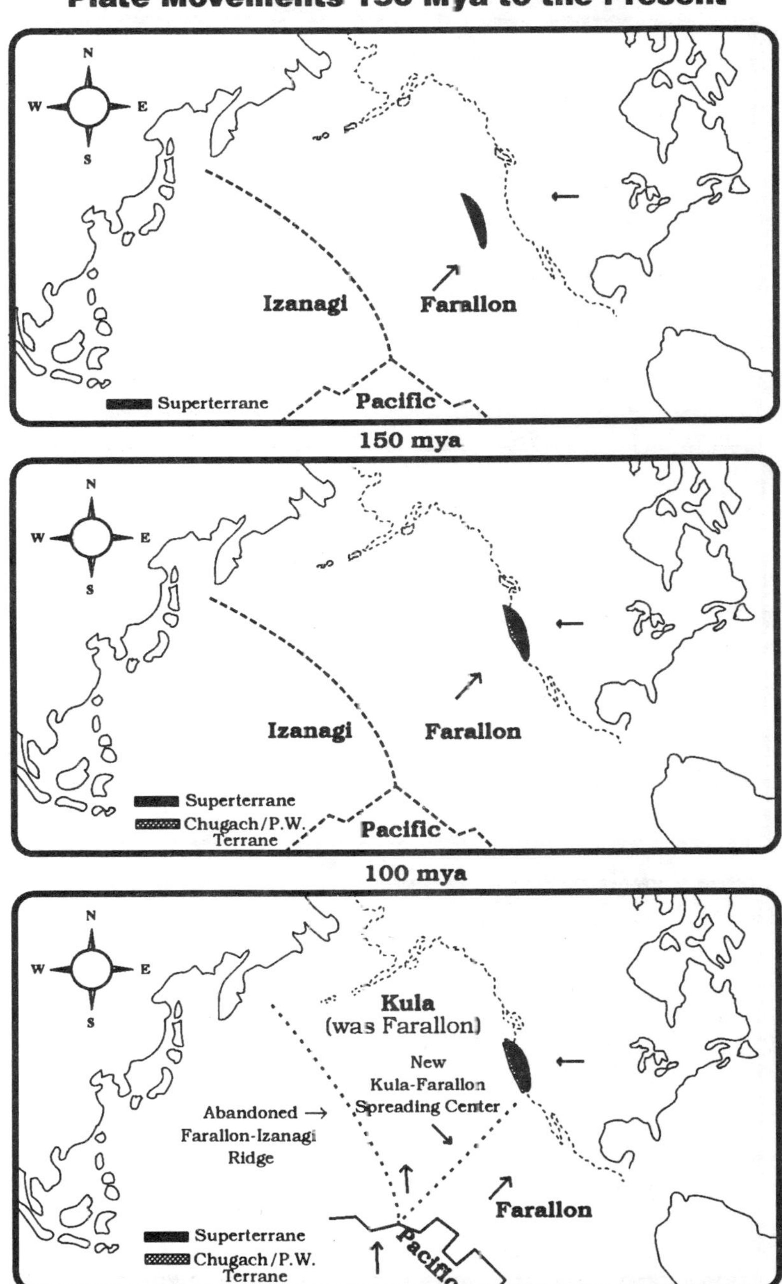

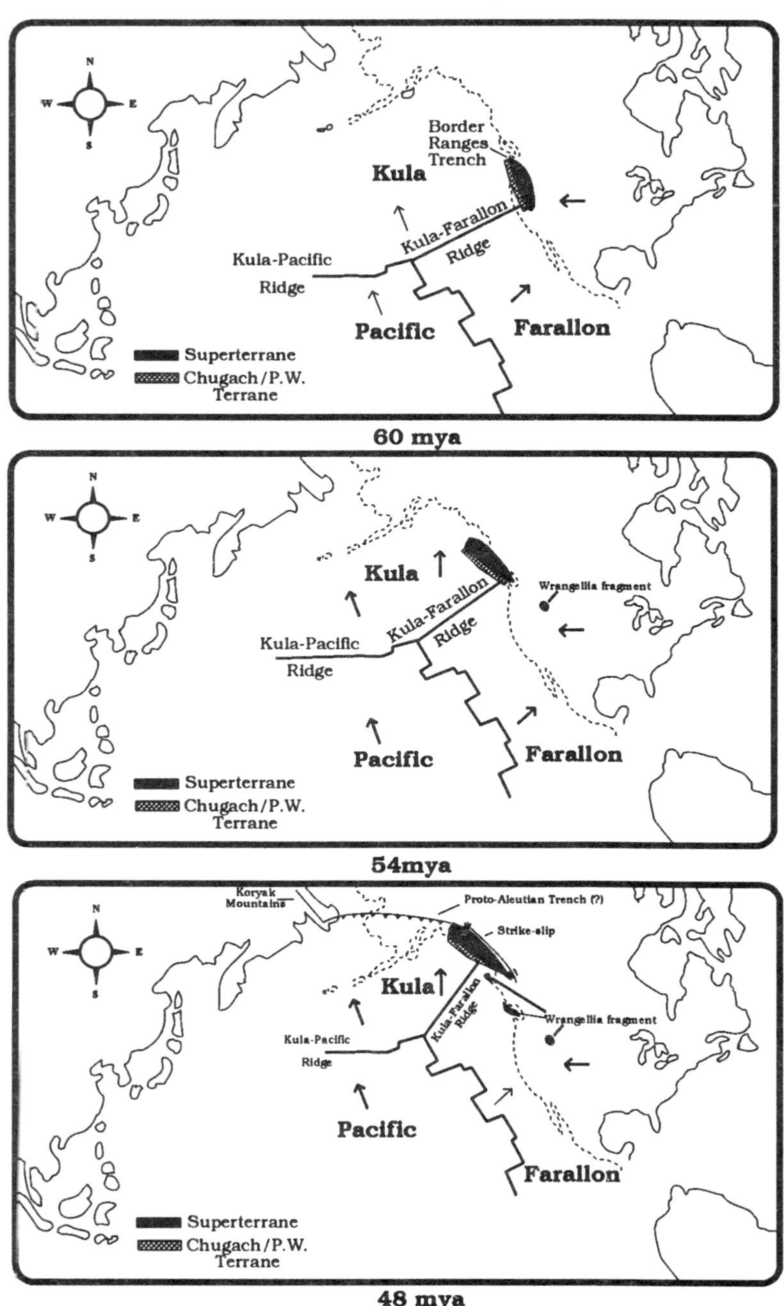
N
W
E
S
Border
Ranges
Trench
Kula
Kula-Pacific
Ridge
Kula-Farallon
Ridge
Pacific
Farallon
Superterrane
Chugach/P.W.
Terrane
60 mya

N
W
E
S
Kula
Kula-Pacific
Ridge
Kula-Farallon
Ridge
Wrangellia fragment
Pacific
Farallon
Superterrane
Chugach/P.W.
Terrane
54mya

N
W
E
S
Koryak
Mountains
Proto-Aleutian Trench (?)
Strike-slip
Kula
Kula-Pacific
Ridge
Kula-Farallon
Ridge
Wrangellia fragment
Pacific
Farallon
Superterrane
Chugach/P.W.
Terrane
48 mya

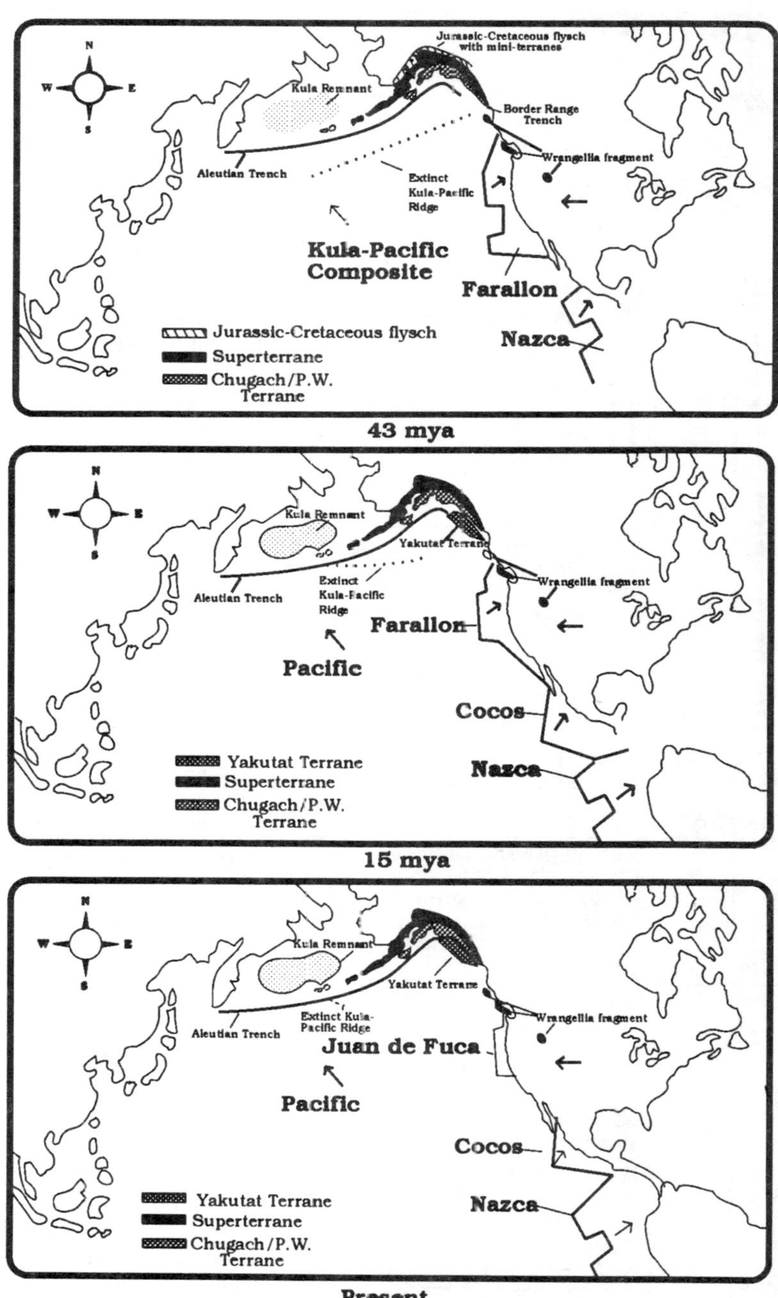

N
W E
S
Jurassic-Cretaceous flysch
with mini-terranes
Kula Remnant
Border Range
Trench
Wrangellia fragment
Aleutian Trench
Extinct
Kula-Pacific
Ridge
Kula-Pacific
Composite
Farallon
Nazca
Jurassic-Cretaceous flysch
Superterrane
Chugach/P.W.
Terrane
43 mya

N
W E
S
Kula Remnant
Yakutat Terrane
Wrangellia fragment
Aleutian Trench
Extinct
Kula-Pacific
Ridge
Farallon
Pacific
Cocos
Nazca
Yakutat Terrane
Superterrane
Chugach/P.W.
Terrane
15 mya

N
W E
S
Kula Remnant
Yakutat Terrane
Wrangellia fragment
Aleutian Trench
Extinct Kula-
Pacific Ridge
Juan de Fuca
Pacific
Cocos
Nazca
Yakutat Terrane
Superterrane
Chugach/P.W.
Terrane
Present

Time Line

Era	Period	Epoch	MYA	Regional Events	Global Events
Paleozoic			500	Alexander Terrane	Early shelled organisms Early fishes Early land plants
Paleozoic	Devonian		408		
Paleozoic	Devonian			Mysterious coral fossils of this age found near Valdez	Numerous corals, fishes, early vertebrates Continental collisions leading to the formation of Pangaea begin.
Paleozoic	Devonian		360		
Paleozoic	Carboniferous	Pennsylvanian		Alexander Terrane	Warm and moist global climate
Paleozoic	Carboniferous	Pennsylvanian	325		
Paleozoic	Carboniferous	Pennsylvanian		Birth of Wrangellian Island Arc near the equator Wrangellia and Alexander terranes joined	Coal swamps, early trees, Insects Amphibians, early reptiles
Paleozoic	Carboniferous	Pennsylvanian	286		
Paleozoic	Permian			Wrangellia uplifted-Permian limestones Earliest rocks of the Peninsular Terrane	Colder, drier global climate World-wide glaciation Final assembly of Pagaea Reptiles prosper First conifers Permian Catastrophe
Paleozoic	Permian		245		
Mesozoic	Triassic			Rifting of Wrangellia - formation of the Nikolai Greenstones Peninsular Terrane and Wrangellia at similar equatorial latitudes	Reptiles dominant First, early dinosaurs Conifers and Palms Early ammonites The Norian Event
Mesozoic	Triassic		208		
Mesozoic	Jurassic			Arc magmatism in the Peninsular Terrane Formation of the Alaska-Aluetian Batholith Peninsular Terrane collides with Wrangellia/Alexander Terrane. Superterrane formed. Magmatism ceases in Peninsular Terrane. Commences in Wrangellia. Superterrane approaches N. America. Gravina-Nutzotin Terrane formed.	Mild global climate Pangaea begins to split apart Central Alaska begins to form through terrane accretion Age of the dinosaurs Ammonites prosper Primitive flowering plants Cycad forests. Primitive mammals and birds. Westward Movement of North American Plate
Mesozoic	Jurassic		144		

Time Line

Era	Period	Epoch	MYA	Regional Events	Global Events
Mesozoic	Cretaceous		144	Collision of Superterrane with North America. Reopening of the Border Ranges Trench Rocks of the Chugach Terrane begin accreting in the Border Ranges Trench. Renewal of magmatism in the Superterrane Accretion of Valdez group	Accretion along the Alaskan margin at the Koryak trench Growth of West Coast of North America through Terrane accretion. Laramide Orogeny Major plate reorganization - Birth of the Kula Plate Carnivorous Dinosaurs Spread of flowering plants Ammonites and Inoceramus clams prosper Early modern mammals Cretaceous Terminal event
			66		
Cenozoic	Tertiary	Paleocene		Accretion of the Orca group and the Ghost Rocks Formation in the Border Ranges Trench Kula-Farallon Ridge interacts with the accreting sediments of the Orca Group and the Ghost Rocks Formation in the Border Ranges Trench. Probable date of Sound's copper sulfide deposits Resurrection Peninsula Ophiolite and perhaps Knight Island Ophiolite Sanak-Baranof anatectic belt of Plutons begins to form	Extinction of the Dinosaurs Beginning of modern life forms. Domination of Birds mammals. North American Plate and Eurasian Plates collide causing Oroclinal bending of Alaska. Reorganization of plate Boundaries begins
			57		
		Eocene		Collage of terranes strike-slip north against the Continental margin. Magmatism ceases in Superterrane Formation of Coast metamorphic belt Sheep Bay anatectic pluton Shortening of Flysch Basin at leading edge of Super-terrane. Final emplacement of Super-terrane and Chugach/Prince William Terrane at Orocline. Terranes transferred to contin-ent by jump in Aleutian trench. Oroclinal bending of emplaced terranes.	Plate boundaries reorganized. Kula plate rotates counter-clockwise and increases in speed Oblique convergence of Kula Plate and North America Uplift and magmatism In Alaska Range Jump of Koryak trench to present position of Aleutian trench Spreading ceases on the Kula-Pacific Ridge - formation of the Kula-Pacific composite plate. Counterclockwise rotation and slowing to present Pacific Plate velocity.
			43		

Time Line

Era	Period	Epoch	MYA	Regional Events	Global Events
Cenozoic	Tertiary	Eocene	43 37	Resumption of Magmatism in Superterrane after 13 m.y. lull during strike-slipping Formation of Aleutian Arc Montague Belt accretes Oroclinal bending of docked terranes	Emergence of modern mammals. Monkeys evolve. Early horses, bats and whales
		Oligocene	24	Plutonism in Northwestern Prince William Sound Formation of the Port Wells Gold District Continued accretion of the Sitkalidak Formation on Kodiak	Tektites formed by Eocene terminal event
		Miocene	5.3	Arrival of Yakutat Terrane at the Aleutian Trench Wrangell Volcanoes Regional uplift Deformation of Orca rocks Formation of Valdez Gold district Mountain Glaciers	Evolution of grass Herbivores prosper Early Apes
		Pliocene	2	Continued Regional Uplift	Australopithicus
		Pleistocene	.012	Except for interglacial periods ice covers area. Major glacial sculpting of the Sound's landscape.	Global temperatures fall Polar ice caps form Mammoths, Saber-toothed tigers. Early Man. Sea levels fall. The Bering Land Bridge forms.
		Holocene	0	Glaciers retreat rapidly. Sea invades the Sound's deep glacially carved basin. Sea levels rise and fiords form. Final Glacial sculpting.	General warming. Isostatic rebound. Sea levels rise. Modern man.

Chapter 1. The Earth in Motion

"Nothing endures but change." Heraclitus

The high afternoon sun beats down warming the small, rounded rocks of the Nellie Juan moraine lulling me into a state of dull and drowsy reverie. The reflected warmth, unlikely for this latitude, suddenly retreats before a chilled zephyr issuing down from the deep-blue face of nearby Nellie Juan Glacier. I start from my afternoon's reverie, suddenly awakened to a renewed awareness of the spectacle that surrounds me. Before me, the iceberg-strewn lagoon gives way to precipitous, gray-granite cliffs wedged apart by the aquamarine, tumbling form of Nellie Juan Glacier. Occasionally, a thunderous rumbling fills the air as mammoth chunks of ice crash from the 200 ft. face into the waters below.

Immediately to the left of the glacier's front lies a stark landscape of gray granite rubble backed by sharp sedimentary peaks rising above a gray granite base. Here, the glacier's rapid retreat has been so recent that soil adequate for plant growth has not yet had an opportunity to form. Farther back to my left, the jagged lunar landscape slowly gives way to a more verdant, rounded scene where mosses and lichens slowly transform the mineral-rich rock into a thin layer of topsoil which

Fig-1. Nellie Juan Glacier from the spit formed by the 1935 moraine.

will soon support the roots of willow, alder, and eventually spruce and hemlock forests. This is a hobbit land of lush, tarn-dotted meadows stretching their green forms over a rolling landscape of glacially scoured granite.

To my right, in the next valley, towering granitic domes dominate the shores of Deep Water Bay — a miniature marine Yosemite. Behind me, stretch the sparkling blue waters of Prince William Sound whose rugged mountains, deeply indented shorelines and forested islands contain innumerable scenes of equal beauty making this one of Earth's truly unique places.

The thought strikes me, as it often does, that I am privileged to be here at this exact point in time and space. Yet, it was not always like this — Prince William Sound. As the sun-warmed sediments filter through the fingers of my outstretched hand, I reflect that even the minerals composing this sand probably originated 200 million years ago somewhere off the ancient equator. How many times had they been molded into solid rock only to be weathered or ground once again into fine sediments before reaching this snapshot in space and time? The thought strikes me that Prince William Sound like the Earth and the Universe itself has been a place of constant change and has endured endless cycles of restless creation and destruction.

Geologically, the forces of creation arise primarily from the Earth's internally produced heat resulting from radioactive decay deep in its crust and molten core. Internal heat provides the motive force that drives the restless crustal plates over the face of the globe. Where plates tear apart in mid-ocean, the crust weakens allowing molten rock to flow forth onto the surface, creating new ocean floor. Colliding plates buckle the earth's crust thrusting up great mountain ranges. The forces of collision melt the deeply buried rock which bursts forth in volcanic violence, spewing out new continental crust or slowly simmers below the surface eventually cooling into great, granitic plutons.

The forces of geologic destruction reside mostly in heat produced externally by the sun. This external heat engine causes the earth's atmosphere to circulate bringing the droughts and dust storms, rains and floods, blizzards and glaciers that erode mountains and send sediments to settle in the deep ocean basins. For billions of years the twin forces of creation and destruction have been embraced in a Shivalike dance whose every movement transforms the face of the Earth. Continents have been born, wandered the face of the Earth and died. Whole oceans have come and gone. Great mountain ranges have thrust skyward only to be relentlessly leveled by the slowly driving forces of erosion.

Evidence for the coming to be and passing away of the Earth's crust abounds from huge, awe-inspiring volcanoes to insignificant sediments washed by raindrops to the sea. But these changes require time — periods of time so vast that even

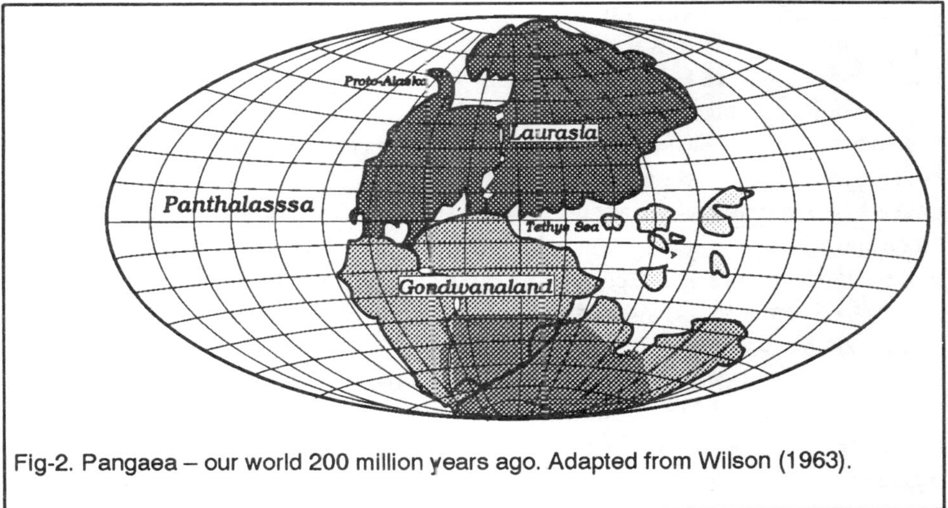

Fig-2. Pangaea — our world 200 million years ago. Adapted from Wilson (1963).

the imagination reels before them. How can a being whose very lifespan is a but a mere flicker in the black abyss of geologic time begin to comprehend the meaning of even a million years? A million years... If we represent this expanse of time by a 12" ruler, the span measuring a human generation would not even be visible to the naked eye. If we represent a million years by a yardstick, then a single life-span would appear as a fine, barely discernible mark. To perceive a single life-time in the expanse of time it took to form the rocks of Prince William Sound would require a measuring stick nearly 90 feet long!

Our story begins 200 million years ago — a period of time so vast as to be almost incomprehensible. Yet, as the age of the earth goes, we are speaking of rather recent events. If we imagine Earth's 4.5 billion years collapsed into the span of a hundred years, then we are considering only the last four and a half years.

Continental Drift

Two hundred million years ago, there was no Prince William Sound, no Alaska and no North America. Instead, all of the dry land masses of the earth were clustered together into one giant super-continent which scientists call *Pangaea* (Greek, *Pan* = all, *Gaea* = Earth). Around Pangaea seethed the restless ancestral Pacific Ocean — *Panthalassa* (all oceans).

To those of us accustomed to seeing the continents in their present position, the

Rand McNally of 200 million years ago would have been a strange sight indeed. The equator slices across the proto-North American continent somewhere in the vicinity of New York City. Capetown and Calcutta are located near the present site of Antarctica. A close examination of the west coast of America reveals that California, Oregon, Washington and most of British Columbia are missing. Where Alaska joins the continent today, a narrow sickle of land, itself only recently arrived, arcs out across the ancient ocean patiently awaiting the arrival of bits and pieces of other continents, submarine plateaus, seamounts, and island arcs which will soon be arriving on the conveyor belt of the Pacific Plates. Over millions upon millions of years, it will slowly gather these to itself to finally assume its present size and shape.

Then, the land was distributed equally between the two hemispheres; whereas today, two-thirds of the continental land masses are located in the northern hemisphere. Much of the Earth's landmass like her later population must have migrated northward. Along the superimposed boundaries of the present continents, one can already perceive linear fractures, some filled by chains of lakes and inland seas — a sign that rift valleys are already forming along lines of weakness in the Earth's crust. Already, processes are beginning which will soon tear the supercontinent asunder. Along the east coast an island-strewn, ancient sea stabs deep into the megacontinent. Scientists have dubbed this equatorial sea the "Tethys Sea" after Tethys, the Titaness who was wife-daughter of the Greek God — Oceanus. Some of the islands of this ancient sea will later collide with the Asian landmass to become Kamchatka, Korea, and Southeast Asia. Others will be plastered up against the west coast of North America leaving fossil assemblages in the Pacific Northwest exhibiting a distinctly equatorial and oriental flavor. The Tethys Sea itself caught between the pincers of rotating continental masses will eventually become trapped into the great inland sea now known as the Mediterranean.

If one follows the contour of the rift at the head of the Tethys Sea around the northern tip of Africa which is pressed close against the east coast of the United States, then crosses the ancient inland sea which will become the Gulf of Mexico and describes a line that cleanly divides north from south America, he has traced out a line where the two ancient hemispheres would soon rift apart. The northern hemisphere, consisting of what would become North America, Europe and Asia, is referred to as *"Laurasia,"* and the southern hemisphere consisting of primitive South America, Africa, Arabia, India, Australia, and Antarctica is called *"Gondwanaland."*

Three decades ago, to assert that the solid earth beneath our feet was not fixed and stable but wandered aimlessly about the globe was tantamount to scientific heresy. Yet, the evidence was there in plain sight. Haphazardly pointing paleomagnetic compasses, equatorial fossils and coal fields in Alaska, oil seepages in high latitudes, glacial striations in the Sahara Desert, and marine fossils atop Mount Everest all went begging for explanations.

Fossil evidence spanning long periods of time indicated that identical plant and animal species inhabited continents now separated by vast expanses of ocean. Paleontologists devised all manner of complicated theories of landbridges, rafts and ancient currents to explain these occurrences; yet the explanation turned out to be quite simple. The reason the same dinosaurs showed up simultaneously in North America and Europe was simple: they walked back and forth between these continents — and not clear across Asia and then over some ancient form of the Bering Land Bridge but across a narrow rift valley that would someday spread so wide as to become the Atlantic Ocean itself. The reason the same species of ancient fern thrived at the same time on the east coast of South America and the west coast of Africa was that these plants once grew on opposite sides of a rift valley not on opposite sides of the Atlantic Ocean. The reason the same ancient crocodiles can be found in both South America and Africa was not because some South American crocodiles climbed aboard some primeval fallen trees to be rafted by ancestral currents to the African continent. Instead, they were rafted there aboard the continents themselves.

Certainly, the idea that continents move had occurred to several thinkers inhabiting the fringes of the orthodoxies of their ages. In 1620 the English Philosopher Francis Bacon noted how the coastlines of South America, North America, Africa and Europe could be made to fit together like pieces of a gigantic jigsaw puzzle. Some 200 years later, the 19th century scientist, Antonio Snyder, noting the similarity between American and European Carboniferous plant fossils asserted that at one time all of the assorted continents had been one. But it was up to the German meteorologist, Alfred Wegner and the South African geologist Alexander du Toit in the 1920s and 1930s to put forth the compelling evidence that the continents had indeed at one time been joined and had subsequently drifted apart. By matching similar geological sequences of the same ages on both sides of the Atlantic, noting similar fossil records in diverse continental locales, and showing that similarly aged glacial till deposits in South America, Africa, Australia and India could be understood only if these continents had been joined together in the vicinity of the South Pole (cf. the shaded area of Gondwanaland above), Wegner and Du Toit argued convincingly for continental drift. Yet, few believed them. Like most scientific revolutions, continental drift was slow in gaining acceptance, mainly because it contradicts the sacred, common sense notions we hold concerning our world. The stumbling block for their theory turned out to be that no one could imagine how the continents like ships on the sea could plow through solid ocean floor. Now 50 years later, we are beginning to understand the mechanism which make these movements possible. This new theory is known as plate tectonics. A knowledge of this new science is essential to an understanding of the geology and geological history of Southern Alaska and Prince William Sound.

Plate Tectonics:

To understand how the apparently fixed and solid earth beneath our feet is moving and has moved for past eons, we must first make an imaginary, Jules Verne-like voyage to the center of the Earth. We will begin our first descent on land beneath the continental crust of Prince William Sound. Our space-age elevator/laboratory is well insulated for we know that the temperatures will rise roughly 60° F. for every halfmile we descend into the crust. A port will allow us to view the various layers while a mechanical arm will retrieve samples for analysis in our portable lab.

The first layer we perceive on our descent consists of fine sediments, tiny loose particles of mud, clay, sand

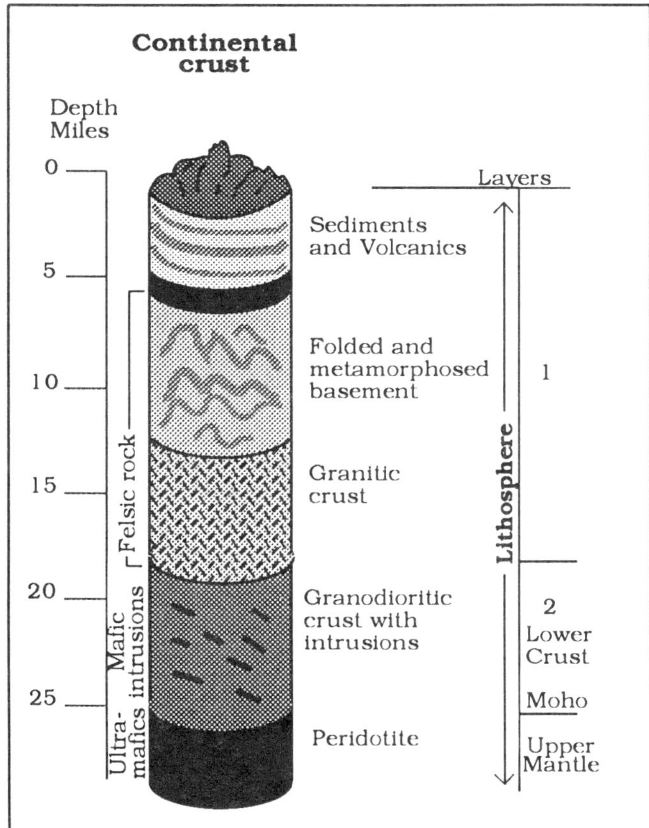

Fig-3. Continental crust. Adapted from Cattermole and Moore (1985).

and volcanic ash which have been deposited from the surrounding mountains by glaciers and other forms of weathering. As we plunge deeper into the Earth's crust, we notice that the sediments gradually give way to solid rock. Apparently, the tremendous weight of the thick overburden of sediments have compressed the muds, clays and sands into solid stone — known as *"sedimentary rock."* These rocks are laid down in orderly, horizontal beds of varying thicknesses and compositions reflecting differing depositional histories. In places the strata are tilted at various angles. In the upper strata we note large cracks known as *"faults"* crosscutting the bedding planes where some awesome force shattered the brittle rock causing it to shear along zones of weakness. However, as we descend deeper where the temperatures and pressures are greater, we make a startling discovery — the rock no longer responds by shattering. Instead the strata become folded into gigantic parallel waves — as if some perverse giant had kneaded the rocks like pliable toffee. Rock at these temperatures and pressures has become more plastic than brittle. We are observing the faulted and folded strata that underlie virtually every continental mountain range.

Examining the lower, folded rocks more carefully, we note that they are of the same chemical composition of those above but the tremendous heat and pressure that folded them has caused them to recrystallize into a wholly new *"metamorphic (changed) rock."* For example, the sedimentary rock, *shale*, compressed from muddy sediments, has become metamorphosed into *slate* and the limestone into marble. Like the sedimentary rocks above, many of these rocks prove to be rich in the oxides of the lighter elements, calcium, sodium, potassium, aluminum and silicon. Such rocks are known to geologists as *"felsic"* rocks because they are rich in *feldspar* (a complex aluminum silicate) and *quartz* (*silica*). The heat and pressure have caused the silica compounds to recrystallize into long, thin, parallel, wavy bands. Nearby, hot, pressurized silica solution has been injected into fissures in the surrounding sedimentary rock to form familiar quartz veins. Samples of these newly formed rocks often fracture cleanly along planes where the new crystals have formed.

Descending farther into the lithosphere (rocky sphere), we encounter a new lighter, unbanded rock layer which like those above is high in silica. We recognize this rock to be common *granite*. What distinguishes these rocks from those above is their matrix of large, interlocking crystals. Every high school chemistry student who has ever patiently grown giant crystals in a petri dish knows that the longer it takes for a crystal to form, the larger the crystal. We can conclude that the large crystals in the granite are a result of its slow cooling beneath the Earth's surface. We know that if the same magma had suddenly erupted forth to rapidly cool in the air, the result would have been a fine grained lava known as "rhyolite." The crystals in rhyolite are so small as to be barely discernible. Extremely rapid cooling results in a glassy volcanic rock known as obsidian. Rocks crystallized from a molten source (magma) are called *"igneous rocks."* Igneous rocks which have cooled slowly beneath the Earth's surface after intruding other *"country rock"* are known as *"intrusives"* or as *"plutonic rock."*

As we descend to 18 miles below the surface, the granitic rock becomes slightly darker and lower in silica having an average composition of *"granodiorite."* When extruded and cooled in the atmosphere, the fine grained, lava formed from this rock is called *"andesite."* Andesitic lavas exhibiting a high water content are characteristic of explosive volcanoes forming on island arcs and at continental margins. Lower down, we notice intrusions of an even darker, coarse grained, igneous rock called *"gabbro."* This denser, rock proves to be lower still in silica and aluminum but rich in the heavier elements, iron and magnesium. Such rocks are termed *"mafic"* rocks (from <u>ma</u>gnesium and <u>ferric</u> [iron]). When extruded and rapidly cooled, gabbro becomes a dark, heavy, fine-grained rock known as *basalt.* Whereas, most ocean floor rocks are composed of basalt, rocks composing continental crust are richer in the lighter minerals, silica and aluminum, which because of their low density seem to have collected near the surface like so much slag from the cooling blast furnace — Earth.

While descending, we have been keeping measurements of the density of the various rocks and have arrived at an average density figure of only 2.7 times that of water. At a depth of 32 miles, we reach an abrupt transition zone between the crust and the upper mantle (the Moho). The *mantle*, roughly 1800 miles thick, appears to be composed of the dark, heavy, ultramafic rock, *peridotite*, simmering at temperatures of over 1000 °C. Under certain conditions, peridotite percolating up through the silica and aluminum-rich layers of the crust, can gradually absorb these minerals to crystallize first as gabbro (low silica/aluminum content), then as granodiorite (medium silica/aluminum content), and finally as granite (higher silica/aluminum content). Peridotite and gabbros, under these conditions may also fractionate off lighter minerals forming rocks of more felsic compositions.

At a depth of about 60 miles with the temperature 13 times that of boiling water, the peridotite of the mantle suddenly becomes less sticky (viscous) and loses its resistance to flow. This fluid but solid layer, thought to have the consistency of silly putty, is called the *"asthenosphere."* Roughly 120 miles in thickness, the asthenosphere acts as a lubricating layer between the the overlying mantle and crust (the *lithosphere*) and the denser and more rigid lower mantle allowing rigid fragments of overlying lithosphere, known as *"plates,"* to shift their positions relative to one another. Fragmented segments of crust and underlying mantle (plates) then float like so many jostling rafts on a solid, dense, but fluid asthenosphere.

Below the mantle, lies the outer core. Here, a 1300 mile thick layer of liquid iron surrounds a solid core of iron having a radius roughly of the moon (800 miles). It is the spinning of the earth coupled with the outer core that scientists believe provides the dynamo for generating the Earth's magnetic field. Temperatures at the Earth's core approach 4,000°C.

A descent through the ocean crust would provide an entirely different scene. Let us begin our trip 100 miles offshore in the Gulf of Alaska just beyond the Aleutian trench. Here, after penetrating the inevitable ocean layer, we would once again encounter layers of sediment and sedimentary rocks similar to those encountered on land. However, the sequence and layering of these sediments might prove quite confusing to a land-based geologist accustomed only to the regularity of slow, even deposition on the continents. As we descend, we notice thin sequences of layers of fine muddy silts, clays and sands representing annual depositions. Some of the more angular grains suggest a volcanic source. Suddenly, this orderly sequence is punctuated by a layer many feet thick, the particles of which have arranged themselves according to size — the larger, heavier particles at the bottom grading up to finer at the top. The land-based geologist might recognize the sedimentary rocks compacted from these sediments as typical *graywacke* (dirty sandstone) and *slate* (mudstone) beds, but he would have a hard time accounting for them in terms of his familiar, terrestrial environment. In fact, as late

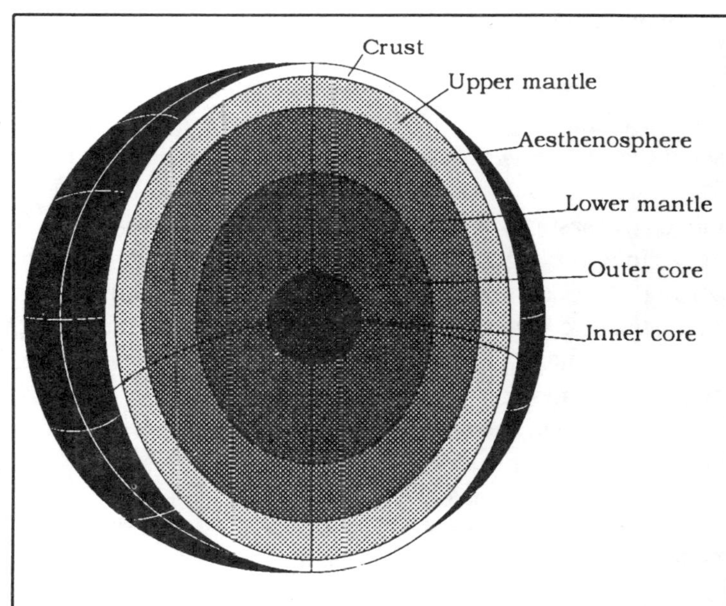

Fig-4. Cross section of Earth showing crust, lithosphere, astehnosphere, mantle, outer and inner cores. Adapted from Cattermole and Moore (1985).

as 1954, geologist, Fred Moffit, examining the rocks of Prince William Sound formed from such sediments remarked, "Without doubt the original muds and sands were derived from a common source; but the causes that brought about the deposition of alternating fine and coarse material, differing so little in composition, are not apparent. The thickness of many beds of slate and graywacke seems to be too great to be accounted for as a result of alternating seasonal changes in erosion and deposition (Moffit,1954)."

Rocks formed from such a sequence are known as *"turbidites"* because they are formed off continental and island arc shelves by *turbidity currents.* A turbidity current occurs when an earthquake, flash flood or volcanic eruption triggers an underwater landslide of loose sediments which have accumulated on unstable slopes at the edge of a shelf or trench. Typically, the turbid, sediment-laden current slices down through a submarine canyon to spread out fan-like on the ocean floor below. So strong are these dense currents that they may transport fine silt up to 400 miles offshore and deposit fans hundreds of feet thick. Tubidite fans resemble the familiar fans formed at the base of snow avalanches. Not all the materials composing these underwater avalanches are fine-grained, however. Large quantities of gritty sand, stream-worn pebbles, cobbles and even occasional small boulders hurtle down the slopes to mingle with the finer sediments below. These are later consolidated into layers of sedimentary rock popularly known as "pudding stone" or "cobblestone" but known to geologists as *"conglomerate."* The pebbles or cobbles in this kind of rock are called *"clasts"* while the finer grained surrounding sandstone is referred to as its *"matrix."*

Under these conditions, we can understand how very thick sediment layers can suddenly interrupt thin and orderly annual sequences. Likewise, we can understand the *graded bedding.* As a turbidity current loses its momentum, it can no longer hold the heavier, larger particles which drop out. These are followed by a rain of increasingly smaller particles until the current in its final death throes has only enough strength to scatter a fine layer of silt over the completed fan. The same sorting also occurs in a horizontal direction across the ocean floor — the lighter sediments are carried farther out on the fan. Because of this kind of sorting, geologists studying turbidites can often infer where particular sequences were deposited in a fan.

If our descent had begun in the murky depths of the abyssal plain far offshore, we would have encountered different sediments. Here, we would encounter a mile-deep layer of siliceous ooze created from a perpetual rain from above of the tiny skeletons of zooplankton. This ooze compacts into a flinty sedimentary rock known as *"chert"* — a favorite aboriginal material for the manufacture of arrowheads.

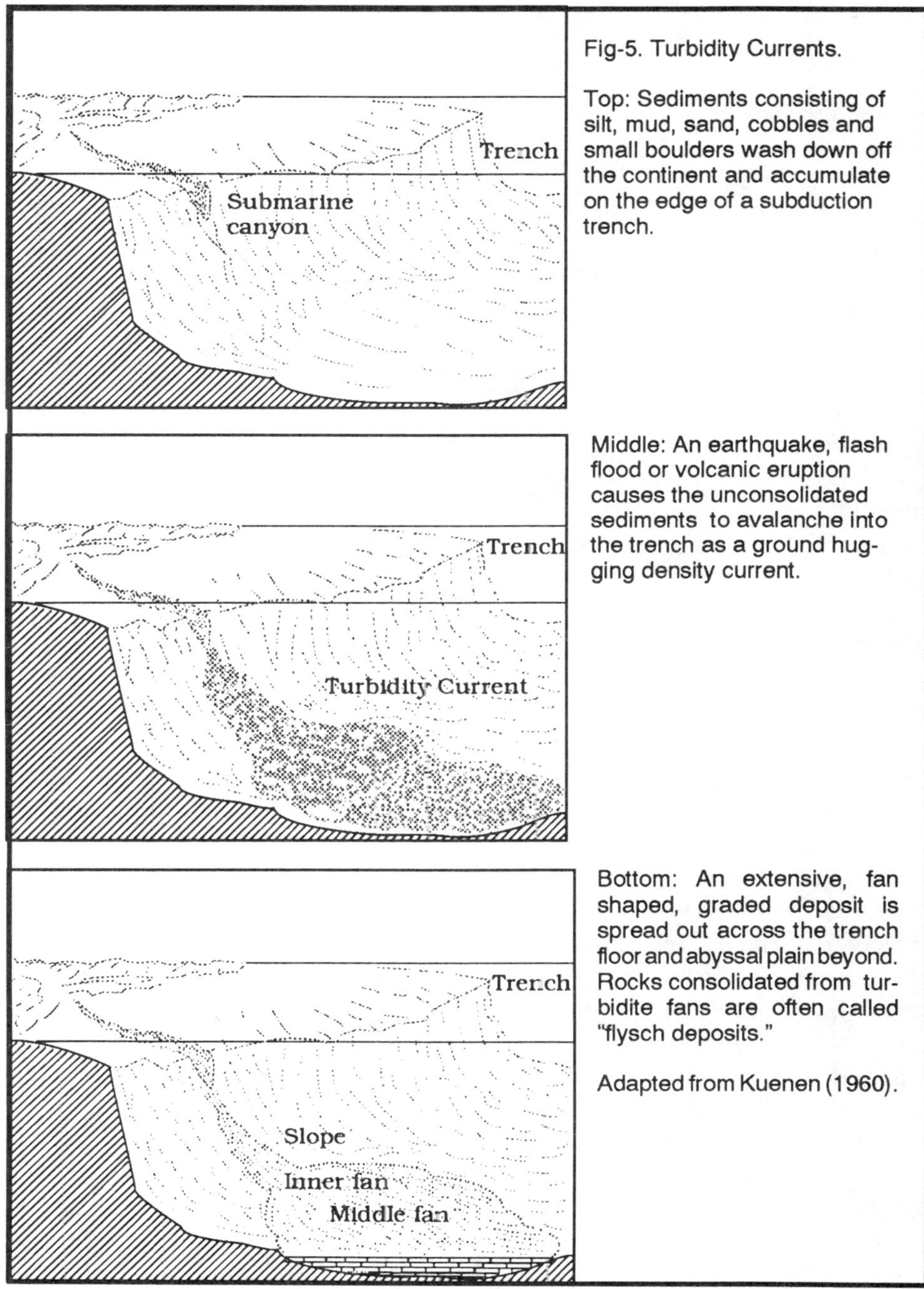

Fig-5. Turbidity Currents.

Top: Sediments consisting of silt, mud, sand, cobbles and small boulders wash down off the continent and accumulate on the edge of a subduction trench.

Middle: An earthquake, flash flood or volcanic eruption causes the unconsolidated sediments to avalanche into the trench as a ground hugging density current.

Bottom: An extensive, fan shaped, graded deposit is spread out across the trench floor and abyssal plain beyond. Rocks consolidated from turbidite fans are often called "flysch deposits."

Adapted from Kuenen (1960).

Fig-6. At low tide, pillow basalts are visible along the shore at Galena Bay.

Below this layer, we would encounter a two mile thick layer of volcanic basalt often in a familiar pillow form. *Pillow basalts* form when hot magma oozes out onto the sea floor. Contact with cold sea water instantly freezes the extruding magma into convoluted surfaces reminiscent of a pile of sand bags. As we might expect, because of the rapid cooling, this basalt exhibits a fine crystalline structure.

Gradually as we descend, we note that the pillow basalts become more and more cut by vertical sheets of basalt. These vertical intrusions are known as *"sheeted dikes"* and represent volcanic feeder vents rising up from the heated bowels of the Earth.

At a depth of scarcely 3 miles, we encounter a three-mile thick layer of gabbro (basaltic magma that has cooled into large crystals beneath the Earth). Chert, pillow basalts, sheeted dikes, and now gabbro — wait a minute. We have seen this sequence before — but on land. Such a sequence on shore would be termed "an ophiolite complex." (cf. fig 10). Knight Island of Prince William Sound and the Resurrection Peninsula near Seward are *ophiolites* and seem to be slivers of ocean crust now emplaced on land. A second oddity strikes us here. We are scarcely three miles below the surface and here we are already encountering gabbro. We are amazed as we did not encounter gabbro intrusions until we were about twenty miles beneath the continental crust. Indeed, the Earth's crust beneath the oceans is very different from that under the continents. We examine the rock samples. The average density appears to be three times that of water whereas rocks from the continental crust were only 2.7 times as dense. Oceanic crust is more dense than continental crust. The ocean rocks are not only more dense, they are darker in color, lower in silica and aluminum and exhibit higher concentrations of the heavier elements, iron and magnesium. Geologists distinguish these rocks as "mafic" or "ultramafic" from the lighter colored and less dense "felsic" rocks of continental crust.

We brace ourselves for new discoveries as we descend ever deeper. Beneath the gabbro layer, we discover peridotite! We must be approaching the mantle. But how could this be? We are only reading a depth of 5 miles. Sure enough, after another mile's descent, we emerge into the Earth's upper mantle. Whereas the mantle lies at a depth of 30 miles or more beneath the surface of continents, we need only penetrate six miles of ocean crust to reach this same layer. From here, we need explore no farther for we know that we will only discover greater pressures and temperatures until we reach the core.

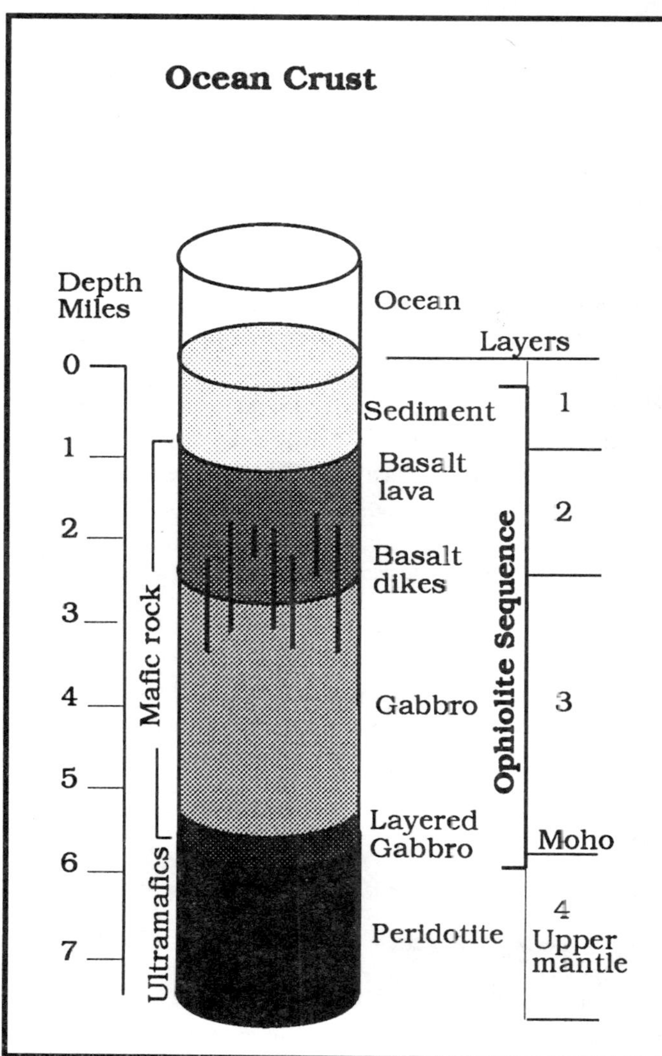

Fig-7. Ocean Crust. Adapted from Cattermole & Moore (1985).

The Restless Plates

Once geologists achieved an understanding of the structure of the Earth and accepted the notion that present land masses occupied different geographic positions in the past, they began to search for a source of energy which could drive the plates across the surface of the globe. Two sources have been suggested — the Earth's heat engine described above and simple gravity.

The Earth's temperature increases with depth. Hence, if the upper mantle has some of the properties of a fluid, we would expect that heated mantle materials at a greater depth would tend to expand. If this expansion due to greater heat is not offset by the pressure at greater depths, then the heated fluid being less dense would tend to rise toward the surface. Nearing the surface, it would gradually cool becoming more dense to sink back into the mantle. Hence, familiar convection currents which makes it possible to boil a cup of coffee or heat an apartment may also account for the movements of continents.

No one knows exactly what causes plate boundaries to form where they do. We suspect that it has something to do with the rising convection currents pictured below. Where two adjacent convection currents rotating in opposite directions reinforce each other, large quantities of molten magma are carried up toward the surface. If the crust is sufficiently thin or weak at this point, an ascending magma plume pierces the surface like a giant cutting torch. Millions of tons of glowing, expanding, molten basalt flow out through this fissure. As the lava pours down the slopes of the ridges on either side, it cools and solidifies. As the basalt cools, it becomes progressively more dense causing the ocean floors on either side of the

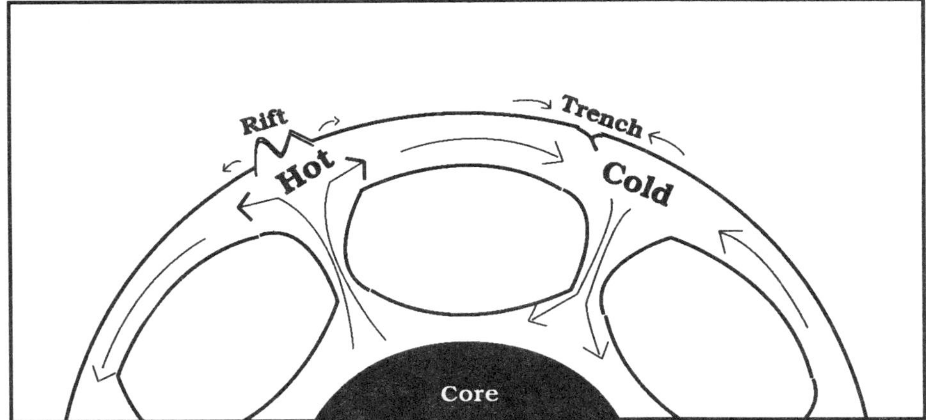

Fig-8. Convective currents created by rising magma from the mantle may drive plate motions.

ridge to subside under their own weight. The ocean in fact becomes steadily deeper in a predictable fashion as we travel progressively farther from a mid-oceanic ridge. Thus a steady downward and outward creep due to the force of gravity probably augments plate motions begun by convection currents. Most plates seem to move apart at the rate of one to two inches a year. However one plate, the Nazca plate, has been clocked at the geologically dizzying rate of over seven inches a year.

Contrasted to most land-based volcanic activity and the violence of island arc volcanic events, the volcanism we have observed on the sea-bottom has been amazingly quiescent. This is mainly due to the composition of the mafic, basaltic lavas which are relatively low in volatiles. Volcanoes on the continental margins and on island arcs tend to extrude explosive, silica rich lavas high in volatiles called "rhyolite" and "andesite."

Seen from a satellite high in orbit, the Earth divested of its oceans, would resemble a gigantic cracked egg. The *mid-oceanic ridge* system creating this impression is a global feature. Stretching for 46,000 miles, this rugged undersea mountain range, averaging 10,000 feet above the ocean floor, girdles the globe. Seen from afar, the entire ridge system has an extremely jagged appearance as if large sections have been offset from a straight ridgeline. At each offset points great cracks called *"transform faults"* extend out thousands of miles perpendicular to the main ridge line. Complimenting this ridge system is an equally impressive system of deep *ocean trenches* — some so profound that a continental feature as

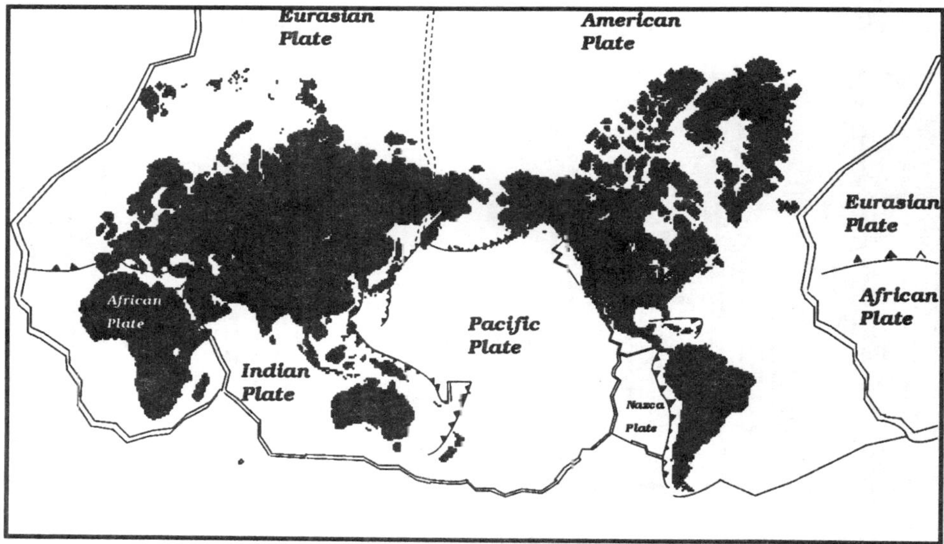

Fig-9. Ridges and trenches define the major plates.

high as Mount Everest could easily be concealed within. Taken together this ridge/ trench system roughly outlines the Earth's plates.

The Earth's surface is made up of roughly, a dozen major plates and several minor plates. These plates are on the average average 75 miles thick beneath the continents and 45 miles thick beneath the oceans. Except for the gigantic Pacific Plate that comprises 1/5 of the Earth's surface area, most of the upper plate layers are composed of both continental and oceanic crust. For example, the North American Plate is composed of Greenland, North America and half the Atlantic ocean. One might say that these continental masses are riding on the plate, and it is plates and not continents that drift.

If we were to descend by submersible to an oceanic spreading zone, we would witness a spectacle few have ever seen. Approaching from afar, we would perceive a high, sinuous, ridge-like mountain range flanked by submarine foothills gradually descending toward the ocean floor. A curious two-mile wide rift runs along the entire ridgeline. Here and there, glowing red mounds of molten basalt dome up from the center of this rifted zone vaporizing the sea water above. On closer examination, the dome is composed of the bulbous extrusions of the pillow basalts we noticed previously on our journey beneath the sea floor. This is the birthplace of the ophiolites.

Following out the cracks of a transform fault, we come upon a field of basalt chimneys spewing forth dense clouds of 300°C, sulfide-rich, black smoke. One of these is over thirty foot high. These submarine hot spring called *"black smokers"* are created by sea water seeping down into the cracks of the weakened sea floor.

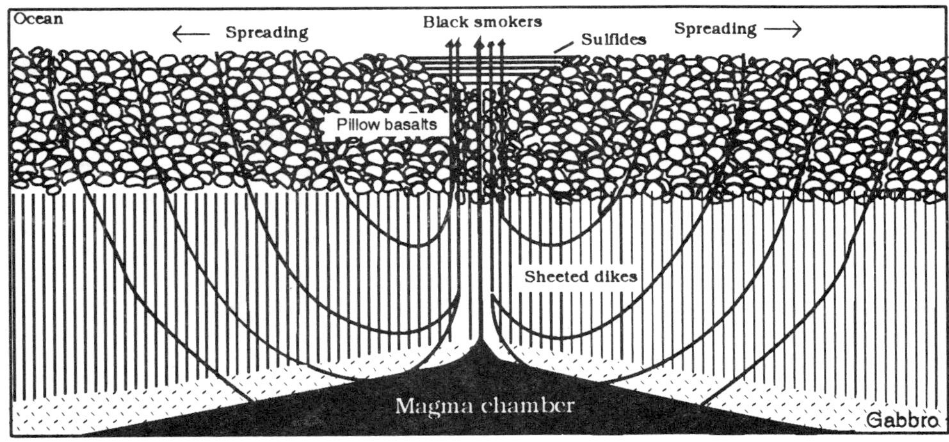

Fig-10. Ophiolite sequences represent new ocean crust created at mid-ocean spreading centers. After Gass (1982).

When the cold sea water encounters the hot, mineral-rich rocks and magma below, it reacts chemically with them. The resulting hot fluid rises through the oceanic crust to spew forth as a black smoke rich in the sulfides of copper, iron, zinc and other metals. If these hot metal concentrates cool and sink into a sea floor depression and are subsequently covered over with a blanket of sediments so that they cannot disperse into the surrounding ocean, they will form a lens-shaped, massive sulfide deposit which may provide a bonanza to some future prospector. If the spreading zone were near a continental margin or off an island arc where turbidity currents frequently avalanche great loads of sediments onto the ocean floor, such a scenario would be even more likely. Under these conditions, a steady rain of heavy sulfides may accumulate into a layer several inches thick until suddenly trapped under a fine layer of cascading sediments. When this process repeats itself numerous times, an ore body of alternating layers of compressed sulfides and narrowly banded sedimentary rock results. Such ore bodies have been discovered at the Midas Mine near Valdez, at Ellamar and on Latouche Island.

Where overlying layers of extruded basalts are penetrated by the throats of rising hot springs, the mineral-rich fluids disperse into the inevitable cracks, crevices, and fissures of the surrounding rock. In this case the cooling fluids create an ore body consisting of a an intricate network of sulfide veins hosted in the basalt like those found at Rua Cove on Knight Island.

Around the black smokers, we note another curious phenomenon. In general, these great depths are devoid of any kind of life, yet clustered around these submarine thermal

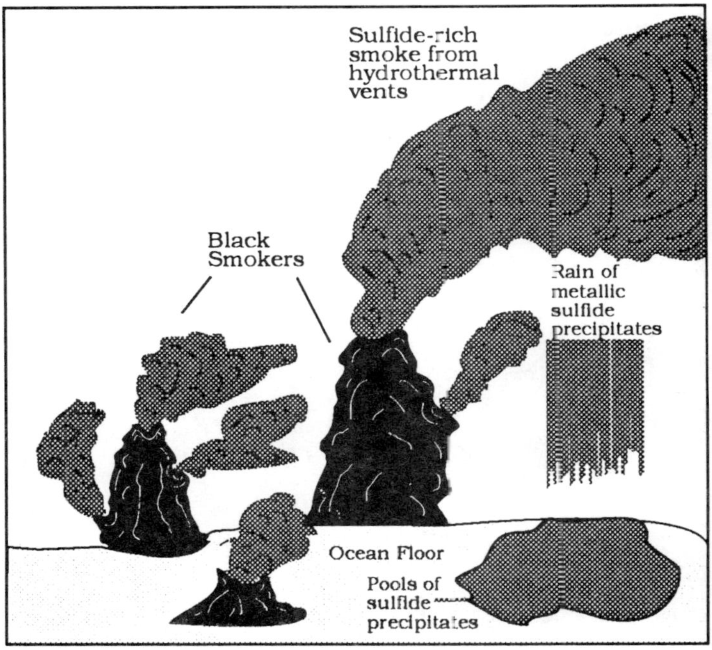

Fig-11. Black smokers are underwater, sulfide-rich, hot springs occuring near spreading centers.

springs, we find giant clams, crabs, and great blood worms. What accounts for this abyssal oasis? What nutrients could possibly sustain this isolated ecosystem? Closer analysis shows that a strange species of bacteria is feeding on the sulfur compounds issuing from the vents providing a first link in the food chain for the larger species.

Before leaving this deep, ocean realm, we will sample and date rocks from the sea bed starting at the mid-ocean ridge traveling outward toward the continental margins. As we would expect, the rocks near the spreading zone are very young indeed. As we move steadily away from the spreading zone, the rocks become progressively older at equal distances from the ridge. This is exactly what we would expect from our sea-floor spreading discovery. However, when we reach the continental margins where the ocean floor should be the oldest, we make an amazing discovery — nowhere can we find a rock older than 180 million years. This is especially astonishing as we know that the oldest rocks found in continental interiors are more than 3,800 million years old. How can it be that continental rocks are so old and oceanic rocks so young? The answer awaits us deep in the ocean trenches.

If we plot all of the earthquakes measured by seismographs around the globe for the years 1961-1967 on a world map and compare it with our map of ocean ridges and trenches above, we make a startling discovery — we have once again sketched out the major outlines of the plate boundaries.

If we distinguish between deep quakes (those occurring several hundred miles down) and shallow (those occurring several miles down), we note that most of the

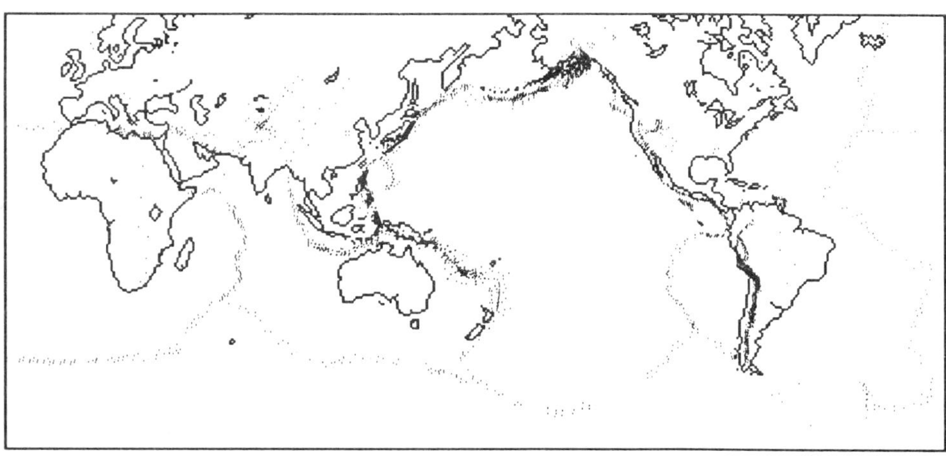

Fig-12. Earthquakes between the years 1961-1967 outline the Earth's plates. After Barazongi and Dorman.

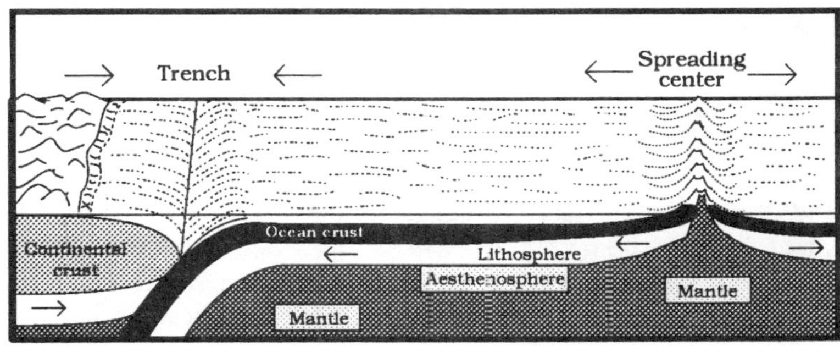

Fig-13. Ocean crust is created in a midocean spreading center and destroyed in a subduction

shallow earthquakes are located along the spreading centers and along transform faults where plates are sliding past each other; whereas most of the deep-seated quakes occur along ocean trenches. Likewise, most andesitic volcanoes, many of them forming island arcs, also seem to parallel the ocean trenches. Most of the world's younger folded and faulted mountain ranges also parallel plate boundaries defined by trenches. These mysterious deep gashes in the sea bed seem to play a major role in deep-seated earthquakes, volcanism, and mountain building; but how do they help to explain the youth of the ocean's floor?

It is obvious that if we have a number of spreading centers where plates are being pulled apart in opposite directions on a finite globe, plate collisions become inevitable. Secondly, if new ocean crust is constantly being created at spreading centers without being destroyed elsewhere, it would follow that the ocean floor would have gradually risen causing the oceans to inundate the continents. The only conclusion we can draw is that ocean floor somewhere is being destroyed. That "somewhere" happens to be deep beneath the ocean trenches. In fact the formation of the ocean trenches themselves is a product of this destructive process.

When a plate composed of dense oceanic crust collides head-on with a plate composed of less dense continental crust, the heavier oceanic plate is forced down underneath the continental plate or conversely the continental plate rides up over the ocean plate. Geologists now recognize that trenches form at this collision site and hence mark a *subduction zone*. They describe the oceanic plate as being subducted beneath the continental plate. At first the oceanic plate dives down at an angle of from 15° to 45°. After a few hundred miles, it dips steeply down. On reaching a depth of about 420 miles it melts to rejoin the mantle.

We can now understand the youthful age of ocean's crust. The limiting age of a particular ocean rock is the time it takes to travel at a rate of several inches a year from its fiery birth at a mid-ocean ridge to its fiery destruction at a trench along a continental margin. A quick calculation of spreading rates and distances involved will show that the upper limits of such a journey is on the order of 200 million years.

The Aleutian Trench and Southern Alaska

Those of us who inhabit Southern Alaska live in a natural geological laboratory where evidence of plate collisions confront us at every turn — from the unpleasant joltings of numerous earthquakes and bothersome dustings of volcanic ash to the spectacular uplift of the snow-capped Chugach range. Quiet reminders of our unique location can be found in great granite peaks, in the folded and faulted strata of the Kenai-Chugach mountains, in numerous, mineral-rich quartz veins, and in the burial of massive sulfide deposits in Prince William Sound; while more violent warnings like the 1964 earthquake and the recent eruptions of Augustine Volcano bear terrible witness that *we are living above a subduction zone!*

Two hundred miles offshore in the Gulf of Alaska beyond Middleton and Kodiak Islands, there lies a sixteen mile wide, deep gash in the floor of the Pacific

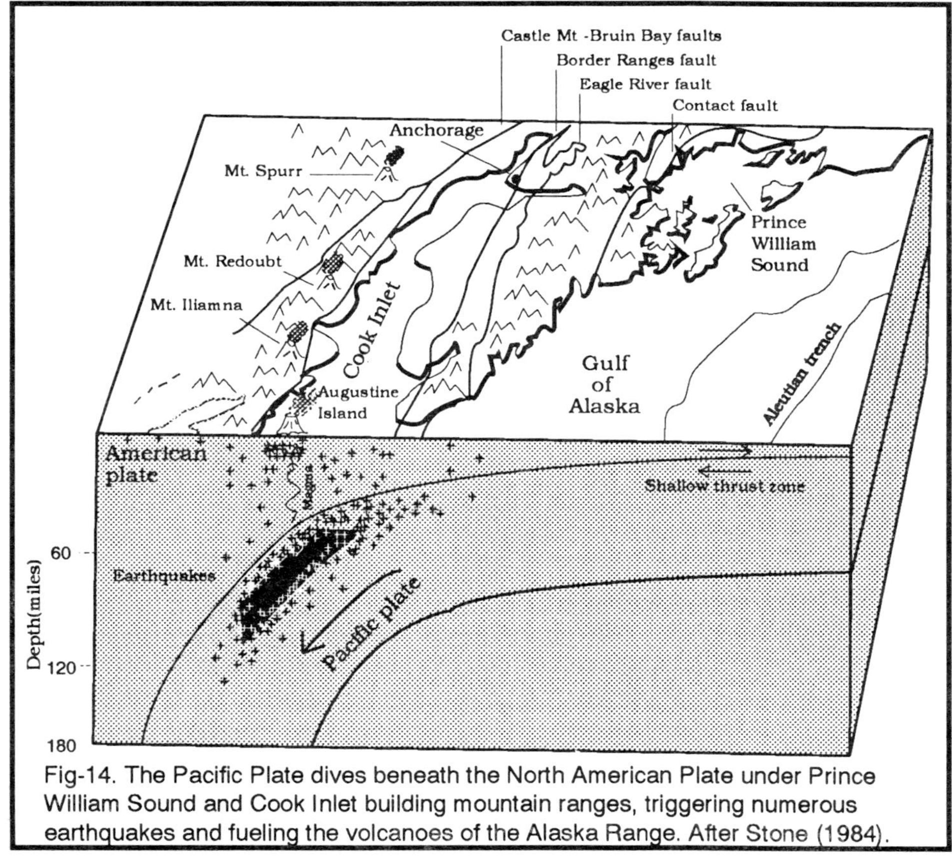

Fig-14. The Pacific Plate dives beneath the North American Plate under Prince William Sound and Cook Inlet building mountain ranges, triggering numerous earthquakes and fueling the volcanoes of the Alaska Range. After Stone (1984).

Ocean known as the *"Aleutian Trench."* Extending from the Prince William Sound area, out beyond the arc of the Aleutian Islands and all the the way to the Kamchatka Peninsula, the 2500 mile long Aleutian trench has been measured to depths greater than 26,500 feet. A one-half mile deep layer of sediments, however, obscures its true depth. It is here that the northwesterly moving Pacific Plate dives down beneath the North American Plate; here ocean crust created millions upon millions of years ago and thousands of miles away at the East Pacific Rise slowly sinks to a fiery death in the Earth's mantle. Granted, the rate of consumption of the Pacific floor here is gradual — only two to three inches a year; however, at this rate, 32 miles of ocean floor disappear every million years. Geologists calculate that since the breakup of Pangaea, two hundred million years ago, an area comparable to that of the entire present floor of the Pacific ocean has been consumed beneath the American Plate.

The most annoying and dangerous consequence of living on the edge of the Aleutian Trench is that subduction zones are notorious for frequent and often violent earthquakes. The Aleutian Arc is one of the most seismically active regions on Earth and the site of numerous devastating earthquakes. These earthquakes are caused by the under-thrusting of the Pacific Plate beneath the Alaskan continental margin. In the Prince William Sound area of the Aleutian Trench, the Pacific Plate dives down at a shallow angle of only 15 degrees as opposed to average angles of 45 degrees (in some trenches ocean plates descend almost vertically). As the plate descends, it begins to bend under its own weight and possibly the compression of collision. The rocks of the cold, descending slab begin to shatter and fault under this stress creating numerous shallow but relatively mild quakes. As the plate thrusts down at a steadily deeper angle to a depth of between 10 and 60 miles, friction builds up on the boundary between the two plates; sometimes the friction is so great that it can temporarily halt plate movements in a given region. When this occurs, tremendous strains build up in the surrounding rocks until suddenly a breaking point is reached; the rocks shatter and the continental plate lurches

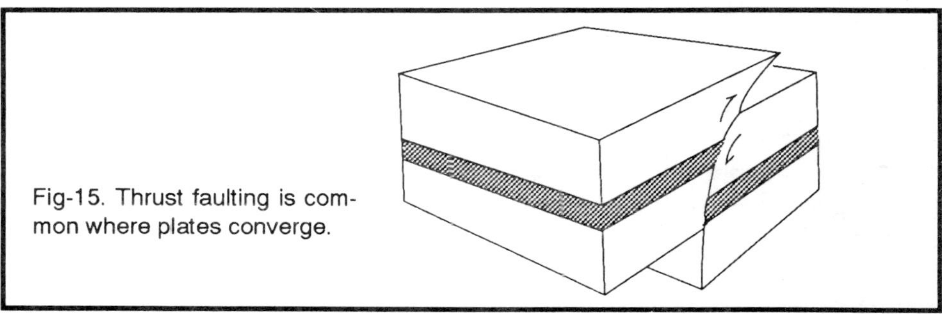

Fig-15. Thrust faulting is common where plates converge.

violently up and over the ocean plate as it did in 1964 and has done on other previous occasions. This kind of tectonic activity produces the strongest quakes and expresses itself at the surface in thrust faulting.

Earthquakes occurring at depths less than 60 miles are classified as shallow quakes. Intermediate quakes occur in the region where the lower plate dives down at an almost vertical angle between 60 and 180 miles beneath the surface. Here, quakes resulting from both compressional and extensional strains in the oceanic plate are less violent; for already the rocks of the colder oceanic plate are beginning to heat and are hence more pliable. The worlds deepest earthquakes occur at depths of from 180 to 420 miles. No earthquakes occur below 420 miles, presumably because the oceanic plate at these depths has melted to rejoined the mantle (Sawkins et. al. 1978).

On the evening of March 27, 1964, some 20 miles beneath the peninsula dividing College Fiord from Unakwik Inlet, frictional stresses in the rocks suddenly gave way and the American Plate lurched up and over the Pacific Plate. One hundred thousand square miles of the Earth's crust were deformed. In Prince William Sound a giant section of continental crust was pivoted up so that portions of Montague Island raised over 35 feet while Harriman Fiord sank 6 feet. The force of the megathrust was so great as to displace the area of Chenega Island 55 feet to the south (Plafker, 1969).

The shallow underthrust angle of the Pacific Plate in the Prince William Sound region probably accounts for the destructive character of the 1964 earthquake. The Sound, although located well back from the Aleutian Trench, sits above the the relatively shallow region where the Pacific plate begins to bend down towards the mantle. At a depth of only 20 miles, the rocks are still relatively cold and brittle. Rather than deforming plastically, they lock up at the plate boundaries creating the stress necessary for a large quake. The sudden release of stress at such a shallow depth releases more energy to the surface than would a more deep-seated quake. The great Alaska earthquake shook with almost twice the energy of the devastating San Francisco quake of 1906.

Those who live above a subduction zone are also destined to suffer periodic volcanic eruptions like those of Katmai in 1912, Mount Spurr in 1953 and 1954, Mount Augustine in 1976 and 1985, and Mt. Redoubt in 1989. As an oceanic plate dives down into a trench, it carries with it a load of silica-rich sediments mixed with sea water. This load of wet sediments begins to heat up under the Earth's natural thermal gradient possibly augmented by heat given off by the forces of friction and compression at the plate boundaries. At a depth of about 70 miles, a complex chemical reaction takes place between the water, sediments and crust; partial melting occurs, resulting in a hot, silica-rich magma which is less dense than the

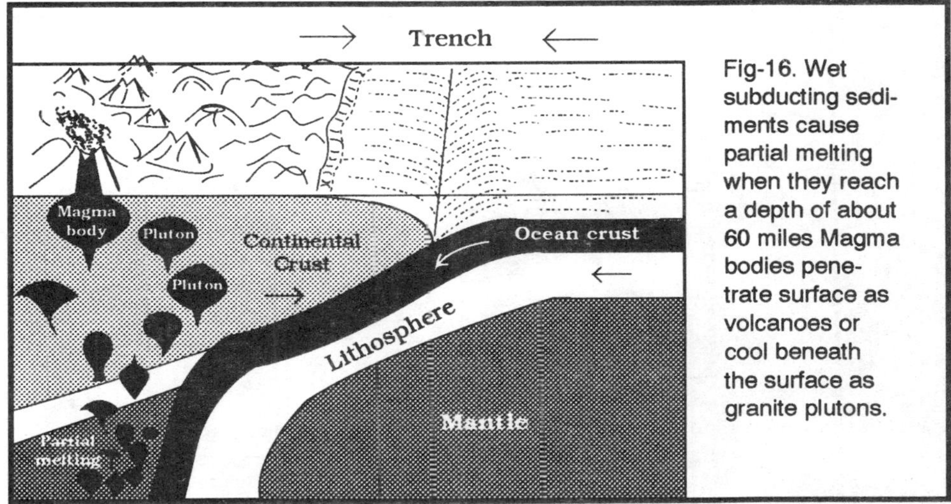

Fig-16. Wet subducting sediments cause partial melting when they reach a depth of about 60 miles Magma bodies penetrate surface as volcanoes or cool beneath the surface as granite plutons.

surrounding mantle. This lighter, molten material begins to rise toward the surface. If it reaches the surface, it expresses itself in an andesitic volcanic eruption like that of the Augustine Volcano. If it cools and crystallizes slowly beneath the surface, it will form a granite pluton.

As a pluton slowly cools, the minerals in the magma body tend to *fractionate* i.e. sort themselves out according to their relative densities — the denser minerals sinking to the interior, the lighter minerals migrating upward. Before cooling, the hot magma body bakes and metamorphoses the surrounding sedimentary rocks to a distance of half a mile or more. Silica-rich, hot quartz solutions bearing fractionated minerals are injected into the surrounding faults, joints, and crevices of metamorphosed sedimentary rock. Gold is sometimes found in such quartz veins.

If the pluton is later uplifted where glaciers are able to scrape off the layers of softer sedimentary rock, it may someday stand out against the sky as a great granite dome like those found in the Nellie Juan region of Prince William Sound.

The Aleutian Volcanic Arc has existed for at least the last 40 million years and presently consists of 76 recognizable volcanoes, 36 of which have been active in historical times. These volcanoes are the result of present subduction of the Pacific Plate beneath the North American Plate at the Aleutian trench. The arcuate shape and andesitic volcanism of the Alaska Peninsula and Aleutian Islands are typical of mid-ocean magmatic arcs where two plates composed of oceanic crust collide in a trench. Typically, the ocean plate which is farthest from its spreading center and hence is older, cooler and denser is subducted beneath the younger, warmer,

Fig-17. The Nellie Juan Pluton towers above Deep Water Bay.

less dense crust. Japan, New Zealand, and the Phillipines are other examples of magmatic island arcs. Because the uplifted, explosive, andesitic volcanoes are composed of easily eroded ashes and lavas, trenches off island arcs are often filled to a great depth by sediments deposited by turbidity currents.

If one measures the distance between the Aleutian Trench and the volcanoes of the Aleutian Chain (the *arc-trench gap*), he would discover a gap of only 120 miles. However, from the trench to its volcanic expression inland from the Prince William Sound area (Mount Spurr) the distance is about 250 miles. The reason for this difference is that off the Aleutian chain, the Pacific Plate dives down at a steep angle reaching depths where partial melting can occur relatively close to the trench. In the vicinity of Prince William Sound, the Pacific plate dives beneath the North American plate at the shallow angle of only 15° hence the region above the 60 mile depth where partial melting occurs is farther back from the trench. In fact, to the Northwest of the Sound, between the volcanoes of the Alaska Range and the Wrangell volcanoes, there is a gap of 340 miles where no volcanism occurs at all. The Pacific Plate behind Prince William Sound probably finally rejoins the mantle far inland in the region of Mount McKinley. Because of the shallow angle of

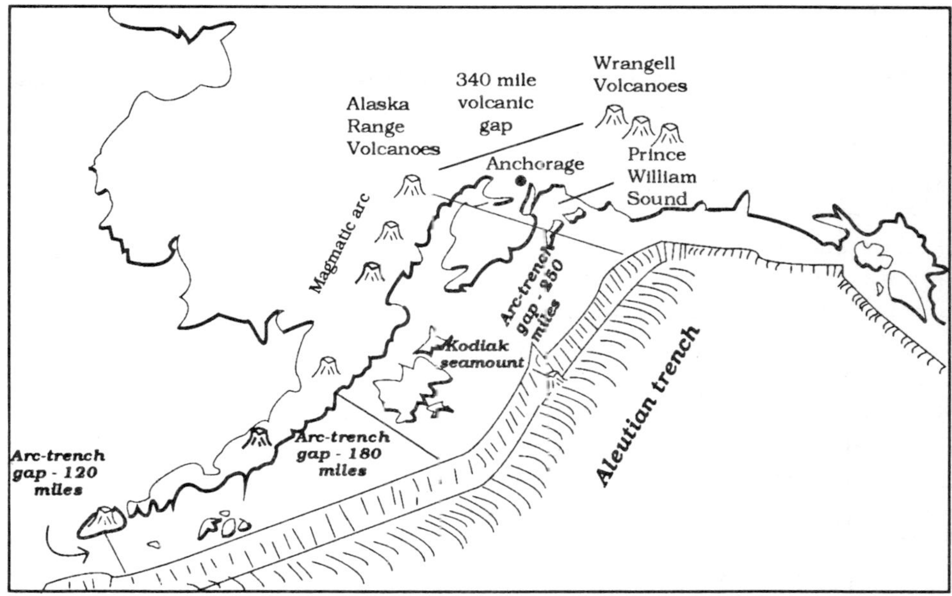

Fig-18. The Aleutian Trench dives steeply down in the Aleutians Islands but exhibits a shallow dip in the Prince William Sound area.

subduction, extremely violent earthquakes will develop beneath the Sound but magma pools are unlikely. Hence the the chances of a volcano or a hot spring emerging in Prince William Sound are remote.

Volcanoes and uplifted, denuded plutons are not the only kind of mountains found at continental margins bordering subduction trenches. From the southern tip of South America to the Chugach range in Alaska, there stretches a vast coastal range of young, folded and faulted mountains. These mountains seem to be thrown up by the tremendous compressive forces at work when two plates collide. Imagine a rug caught in an opening door; under the compressive force, the rug crumples into a series of wave-like folds. Similarly, waves of coastal mountain ranges seem to have been uplifted by plate collisions — certain areas are raised while others are down-warped. The drowned cirques and fiords of Prince William Sound and especially the Kenai Fiords suggest that these areas are in general subsiding, while the coast between Southeast Alaska and Prince William Sound is rising. Here, one finds the loftiest mountains, the Saint Elias range; any boater who has made the trip from Cape Spencer to to Cape Hinchinbrook has bemoaned the lack of convenient anchorages in this area created by the uplift.

The rocks of these coastal ranges also bear witness to their creation at the edge of a subduction zone. A close examination shows that the strata are often tilted at various angles to the horizontal; some strata have been upended to assume an almost vertical orientation. Numerous thrust faults are usually in evidence. The jumbled, overturned, and upended strata are the results of compressive forces at work at the continent's edge. Where the rock has been heated through deep burial, nearness to plutonic intrusions and/or heat given off from frictional and compressive tectonic forces, it becomes more pliable; rather than shattering under stress, it warps into folded strata. Often, if the heat and pressure are great enough, partial or sometimes even total metamorphism of the folded strata will occur. Under these conditions, hot, silica-rich fluids melt out of the rocks to migrate through the cracks and joints of sedimentary rocks where they leach out and concentrate valuable minerals. The gold-bearing quartz veins of the Valdez area are probably due to these processes accompanying uplift.

Subduction trenches near continental margins or next to magmatic island arcs tend to accumulate vast quantities of sediments. The deep-water siliceous ooze which forms chert is brought in slowly but steadily on the conveyor belt of the Pacific Plate. Turbidity currents avalanching down off the continental or arc-shelf spread out across the trenches sometimes furrowing deep canyons into the inner trench slope. Folded and faulted turbidites known as *"flysch"* are common to mountains formed near a subduction trench. Some of these mixed sediments are carried down into the trench on the down-going, ocean plate where they are *lithified* (turned to rock) by heat and pressure. The resulting rock composed of a chaotic jumble of sandstones, mudstones, conglomerates, chert, and fragments of basaltic ocean floor are commonly scraped off onto the underside of the upper continental plate. If subduction ceases and these deep-seated trench fills later become uplifted until they are exposed at the surface, they provide evidence of a former subduction trench. Because of their chaotic composition such trench-fills emplaced on land are called tectonic *"melanges"* (French "mixtures"). The McHugh melange bordering the Seward Highway on Turnagain Arm represents such a fossilized trench.

The *McHugh Complex* as well as the Chugach Mountains as a whole are part of an extensive paired melange-flysch complex, typical of mountains formed above a subduction zone. The paired complex known as the *"Chugach Terrane"* stretches 1200 miles in an arc from Baranof Island in Southeast Alaska to Kodiak and Sanak Islands to the west. Melanges similar to those of the McHugh mark the location of the former trench on Baranof Island in Southeast Alaska and on Kodiak Island to the west. All of these lie along a major fault known as the *"Border Ranges Fault."* We will thenceforth refer to this former trench as the *"Border Ranges Trench."*

Most sediments are not stuffed deeply enough down into the trench to be lithified, deformed and regurgitated as melanges. Instead, because they are too light to be deeply subducted, they pile up in a great wedge against the inner wall (continental side) of the trench. Like a giant snow plow, the continental trench-wall scrapes these wet, unconsolidated sediments off the down-going plate compressing and folding them into solid rock. Some are buried deeply enough to be metamorphosed and scrapped off onto the underside of the continental plate. Later uplifted these appear as a *slate belt* — a group of rocks having a slatey cleavage. Often, lower down in the wedge, a shear zone may develop between the underplating slates and relatively undeformed and coherent beds of turbidites which continue to subduct. This zone of detachment between these two groups is often recognizable on the surface as a melange (Byrne, 1986). Because of the resistance offered the subducting wedge by the continental crust, shear zones known as thrust faults tend to develop in the subducting sediment piles. The constant addition of

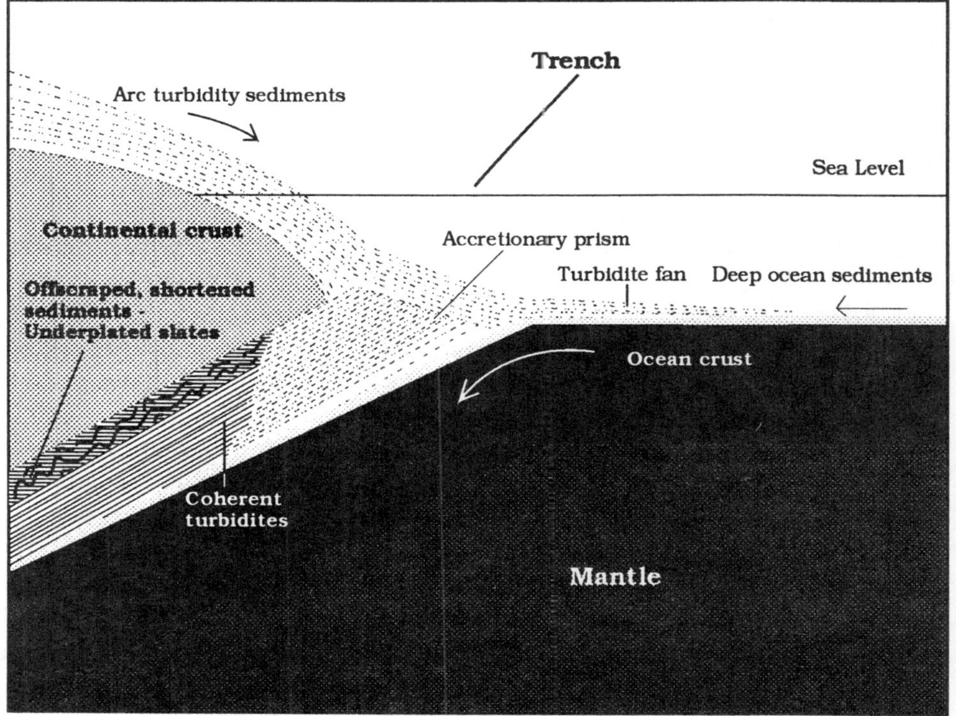

Fig-19. Turbidite sediments pile up against the inner trench wall as an accretionary prism. Subducted sediments are often underplated to the continental plate forming slate belts.

new material to the continental edges by this process is called *"accretion"* and the wedges of material transferred across the trench and partially subducted are referred to as *"accretionary wedges"* or *"accretionary prisms."*

Sufficient heat due to the forces of subduction and deep burial may metamorphose the sedimentary rocks in the upper part of the wedge into *greenschists*. Greenschists form at relatively low pressures and at temperatures of from 300-500 °C. Under these conditions the greenish minerals chlorite and epidote form giving these rocks their characteristic greenish color. Greenschists also form during metamorphism accompanying regional uplift. If temperatures of the wet accreting sediments reach about 1,000 °C, partial melting of a portion of the wedge may occur and huge masses of sediment can fuse to form granite plutons known as *"anatectic plutons."* These plutons may form in the accretionary wedge and differ slightly in their chemistry from plutons fractionated from rising mantle material in a magmatic arc (Hudson et. al. 1979).

The buried root of an accretionary wedge as it is squeezed into the trench is an area of intense metamorphism. Because the ocean plate is relatively cold, the metamorphism in this region is characterized by low temperature and high pressure. Under these conditions, minerals known as *blueschists* form. Blueschists are relatively rare as they form only under a narrow range of conditions where rocks are rapidly shoved into a trench, recrystallized under great pressure and then uplifted to the surface before they can get very hot. Blueschist/greenschist belts like tectonic melanges are signs of former oceanic trenches.

Sediments are not the only passengers on the conveyor belt of the ocean plates. Basaltic Seamounts, submarine plateaus, remnants of mid-ocean ridges, rifted sections of continental crust, and even large magmatic island arcs can be observed moving slowly but relentlessly toward subduction zones. Presently, off Kodiak Island, a seamount is inclined precariously on the edge of the Aleutian trench (cf. Fig-18). These rafted fragments of continental and oceanic crust are called *"terranes."* The term "terrane" should not be confused with "terrain" — a general topological feature. A terrane is defined technically as "a fault-bounded entity, usually of regional extent, that is characterized by a distinctive stratigraphic sequence or rock assemblage that differs markedly from those of nearby , partly or entirely coeval neighbors (Jones et. al. 1981)." Put simply, this definition states that to be classified as a "terrane" a group of rocks must pass three tests: 1) it must be a significantly large group of rocks; 2) it must be separated from adjacent groups by major fault boundaries; and 3) it must differ markedly from these neighboring groups.

When collisions occur between terranes or terranes and a larger continental mass, a couple of scenarios are possible. If the rafted terrane is relatively small such

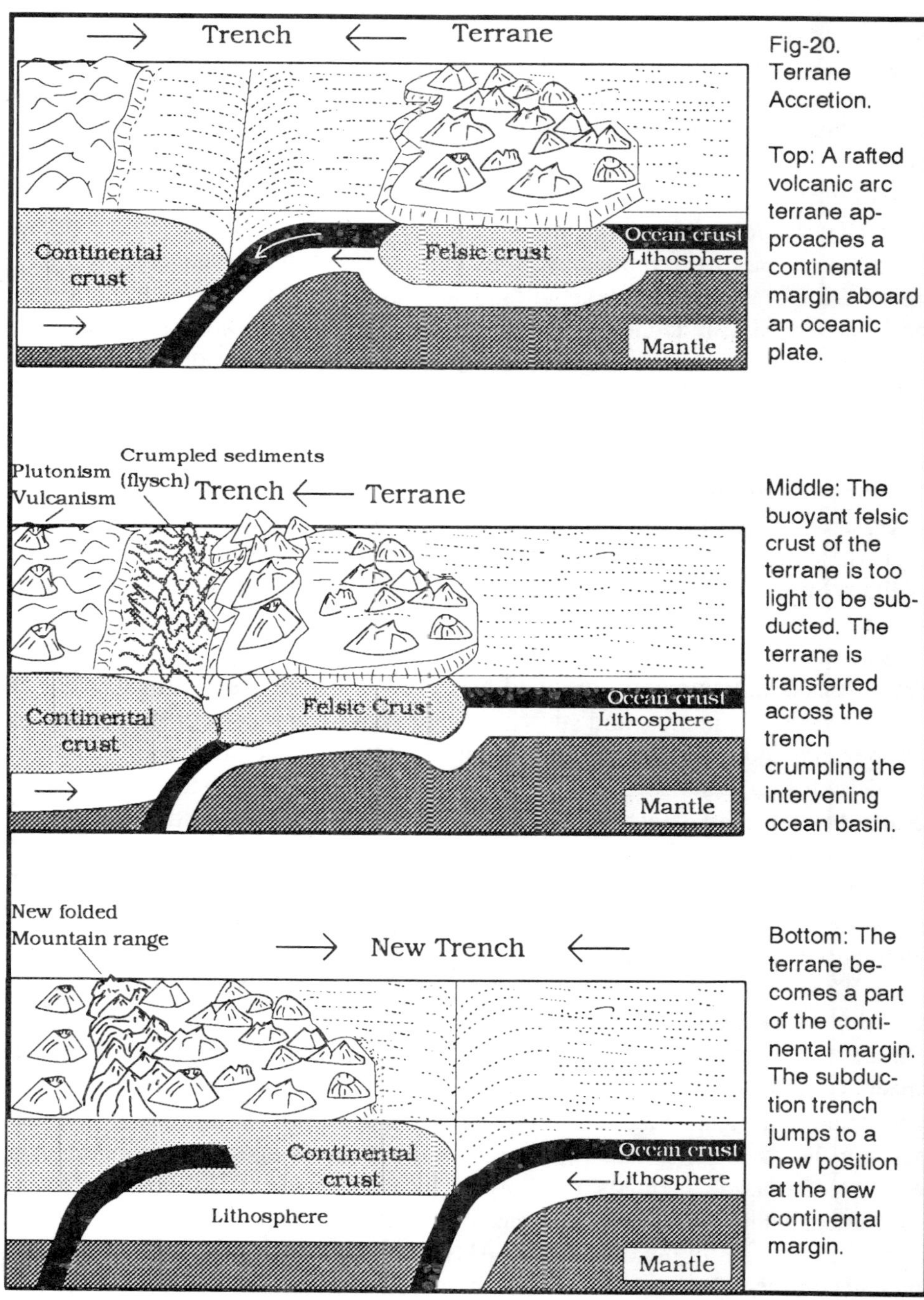

Fig-20. Terrane Accretion.

Top: A rafted volcanic arc terrane approaches a continental margin aboard an oceanic plate.

Middle: The buoyant felsic crust of the terrane is too light to be subducted. The terrane is transferred across the trench crumpling the intervening ocean basin.

Bottom: The terrane becomes a part of the continental margin. The subduction trench jumps to a new position at the new continental margin.

as a lone seamount or small fragment of mid-oceanic ridge (an ophiolite) and its density or geometry is such as to preclude its being shoved down into the trench, then the leading edge of the continental plate may scrape it off the down-going ocean plate. In this manner the block is transferred from the ocean plate to the continental plate in a process geologists refer to as *"obduction"* or *"obductive offscrapng."*

If the rafted terrane is a large micro-continent or great magmatic island arc then the collision may have more disastrous consequences. First, the collision may contribute to dramatic mountain building far inland. Some scientists speculate that the so-called "Larimide Orogeny," which raised the Rocky Mountains about 60 million years ago, was the result of numerous terranes crashing into the west coast of North America. The Alaska Range may have been raised by the collision of a megaterrane which now composes most of Southern Alaska. When such a large terrane docks, it effectively closes an ocean basin and blocks the trench between itself and the continent. As a result, the subduction zone jumps outboard of the newly accreted terrane to take up renewed activity just offshore (Fig-20).

Terrane theory, so necessary to the understanding of the geology of the Prince William Sound area, is of rather recent origin and, in fact, received its major impetus from studies done in Alaska during the 1970s. In an attempt to evaluate the state's mineral resources so that disputes concerning the right-of-way over the Trans-Alaska pipeline corridor could be resolved, geologists swarmed over the state looking for minerals and studying its geology. The results of their research was puzzling indeed; a geologist would piece together a coherent story of the strata in a certain area, walk across a major fault to discover that the strata there told a totally different story. Not only were the rocks of a different kind and sequence, but they were formed in vastly different ages. And what was even more surprising, often they appeared to have been formed in vastly different parts of the globe! Virtually the whole state appears to be made up of a patch-work of accumulated fragments which have been rafted in on ocean plates over the past 160 million years. The ancient ocean plates, precursors of the present Pacific Plate, thought to have deposited these foreign bits and pieces are known as the *"Kula"* and *"Farallon Plates."* A fragment of the Farallon Plate known as the "Juan de Fuca Plate" is presently subducting under the coasts of Washington, Oregon and Northern California. The Kula Plate is thought to have been almost entirely consumed beneath the coast of Alaska by about 40 million years ago. After studying the geology of Alaska, scientists took another look at the rest of North America and concluded that its entire west coast was formed by the accretion of terranes. In fact, it is estimated that in the time it took Alaska to increase from a small triangle of ancient continental crust on the Yukon border to its present size, the western edge of the North American Continent grew by 20% (Jones et al. 1982).

Scientists are able to recognize that the rocks of a specific terrane were formed in a distant location by studying their *"paleomagnetism."* When sediments are laid down or when molten rocks solidify, tiny magnetic mineral grains like minute compass needles in the rock align themselves with the Earth's local magnetic field. If these tiny compasses point away from the present magnetic pole, we can assume that either the rocks have been rotated or that they were formed in a different location on the globe. Even more significant information is conveyed by the dip of these tiny compass needles. Because the lines of force of the Earth's magnetic field converge at the poles, the needle of a compass held parallel to the Earth's surface will exhibit a downward dip whose angle is dependent on one's position on the globe. At the equator where the lines of force are parallel to the Earth's surface the compass needle will exhibit no dip. However, as one approaches the poles, the needle will experience a steadily increasing angle of dip. The correlation between angle of dip and distance from the magnetic pole is so precise that a given dip indicates a precise latitude (but not longitude). Hence, by measuring the dip of the magnetic grains frozen into a rock at its time of formation, geologist can determine the latitude at which the rock was formed.

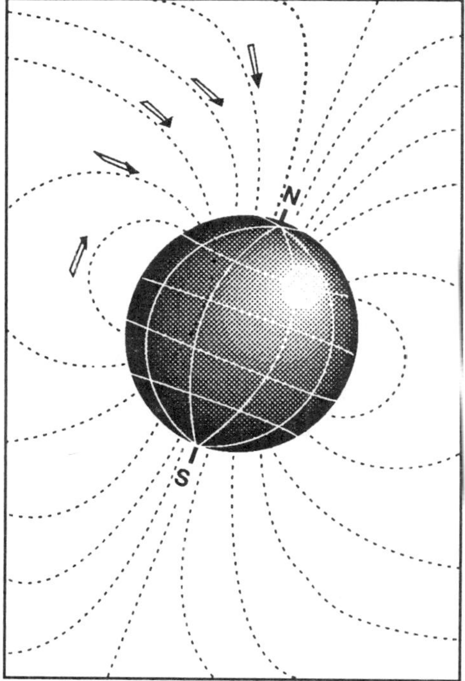

Many geologists feel that because accurate paleomagnetic measurements are difficult to obtain that they should be treated with caution — especially when they appear to contradict more direct forms of geological evidence.

Fig-21. A compass needle exhibits an increasing angle of dip as one approaches the poles. The angle of dip can be used to measure latitude.

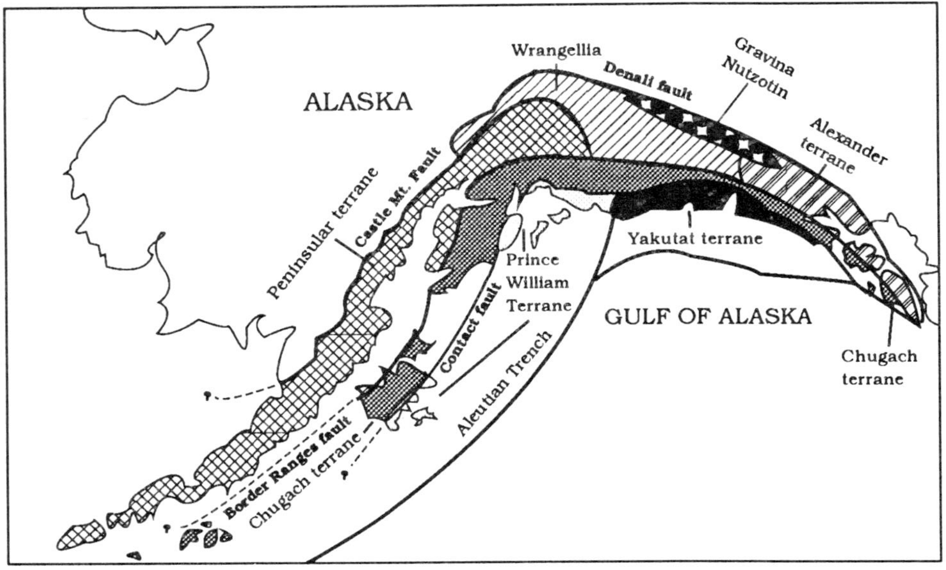

Fig-22. Generalized terrane map of the southern part of Alaska showing Prince William, Chugach, Peninsular, Wrangellia, Gravina Nutzotin, and Yakutat terranes plus the major faults. After Dumoulin (1987).

Terranes of Southern Alaska

Geologists studying Southern Alaska distinguish seven terranes — the Alexander, Wrangellia, Peninsular, Gravina-Nutzotin, Chugach, Prince William, and Yakutat terranes.

Observing the terranes on the above map, one is immediately struck by their peculiar geometry. Virtually all of the terranes seem to have been stretched out along a single axis, and those in the area of Prince William Sound appear to have been bent into elongated arcs. The outboard terranes of Southeastern Alaska and British Columbia also share this same oblong (but not arcuate) geometry.

To understand this peculiar geometry, we must examine the concept of strike-slip or transform faulting. As every school child knows, a slice of Southern California on which Los Angeles rests is slipping northward with respect to the North American Plate along the San Andreas fault causing numerous earthquakes such as the one experienced recently (1989) in the San Francisco area. When two plates meet head on at a trench, the result is thrust faulting; when they collide obliquely, sliding past each other, the result is strike-slip faulting.

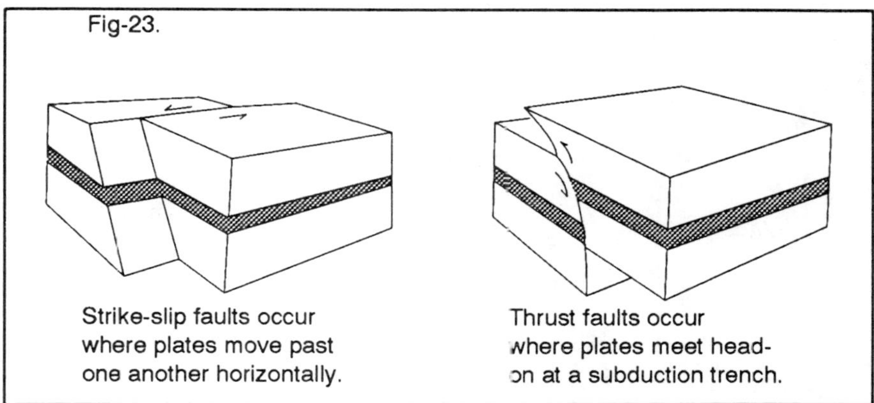

If a terrane riding aboard an ocean plate encounters a continental margin at an oblique angle along a major transform fault zone, strike-slip faults tend to develop in the terrane rending it over a long period of time into a number of elongated crustal slivers. These great slivers of crust are displaced along faults like a deck of cards slid obliquely along a wall. The cards tend to distribute themselves longitudinally, slipping at spaces between each card. The spaces between the cards are analogous to weaknesses in the crust that form the strike-slip faults in the terrane. Geologists refer to this process as *"telescoping"* of the terrane (Jones et. al. 1982).

If this telescoping process occurs over millions upon millions of years fragments of the original terrane may be separated from one another and spread out longitudinally over great

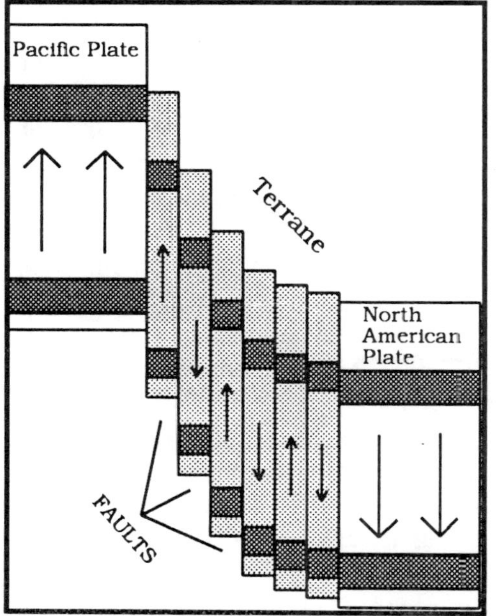

Fig-24. Telescoping. A terrane riding aboard an ocean plate but strike-slipping along a continental margin may be drown out in long slivers along internal strikeslip faults. After Jones (1982).

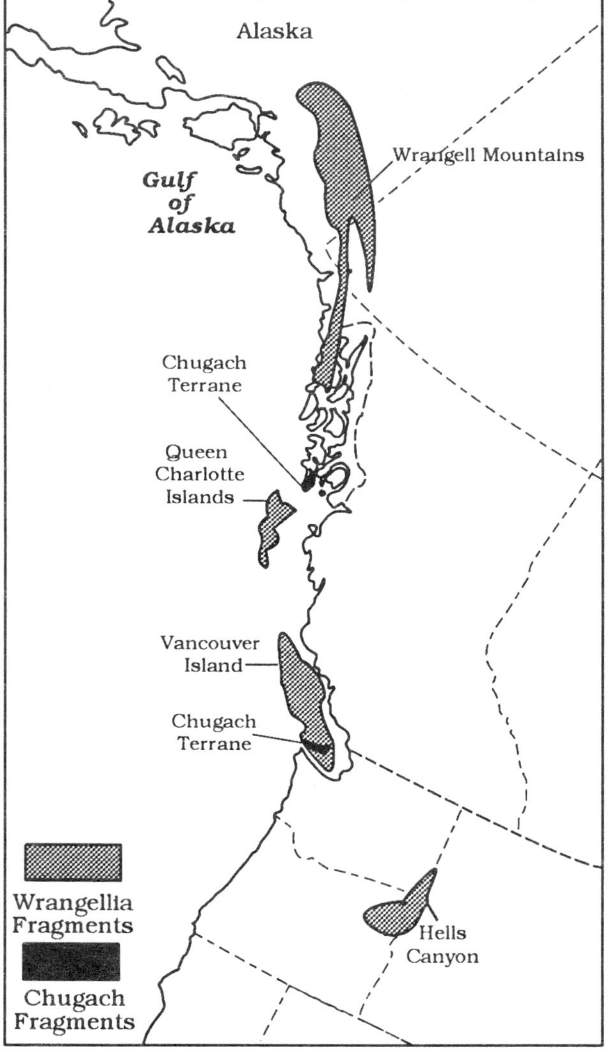

Fig-25. Fragments of the Wrangellia Terrane are elongated and stretch from Oregon to the Wrangell Mountains. Fragments of the Chugach Terrane are found on southern Vancouver Island and just south of Sitka. Adapted from Jones et. al. (1977) and Cowan (1982).

distances. Fragments of the southern Alaska terrane, Wrangellia, occur distributed from southeastern Oregon to the Wrangell Mountains near Glennallen. A small fragment of the Chugach Terrane adhering to a fragment of Wrangellia on southern Vancouver Island seems to match another fragment of the same terrane on the Southern end of Baranof Island 675 miles to the north.

Anyone who has taken the Ferry through the transform fault zone of Southeast Alaska and British Columbia has observed first-hand the results of these geological processes. The route follows long, northwest-southeast trending fiords formed from strike-slip fault zones later gouged out by Pleistocene Glaciers. The shores of these fiords are the elongated, dismembered fragments of various terranes that strike-slipped their way up the coastline tens of millions of years ago.

Numerous Paleomagnetic studies of terranes south of the Denali Fault system

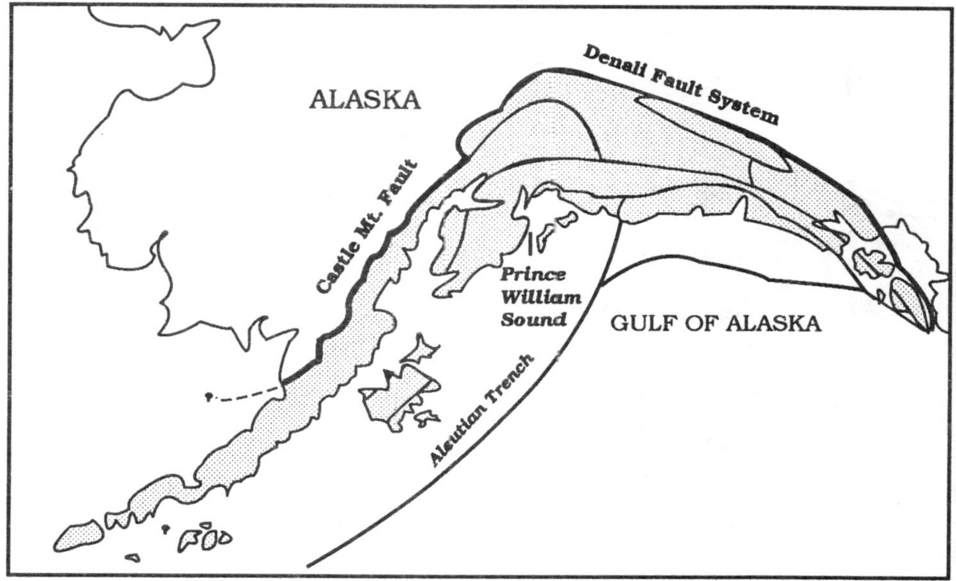

Fig-26. Paleomagnetic evidence suggests that terranes north and west of the Denali-Castle Mt. fault systems seem to have been in place since Cretaceous times; whereas terranes indicated in grey above may have undergone significant northward displacements since the Cretaceous.

and east of the Castle Mountain Fault system seem to agree that they have all undergone significant northward displacement while the terranes of western and central Alaska north and west of the these fault systems seem not to have moved since Cretaceous times (Cole et al. 1985). Most of these terranes, however, have undergone significant counterclockwise rotation.

Geologists speculate that the Border Ranges Trench which separates the Chugach Terrane from the Peninsular, Wrangellia and Alexander terranes may represent a Mesozoic trench itself displaced northward in more recent times by transform faulting. Evidence of major strike-slip activity on this fault was noted by Pavlis (1982). Scientists assume that a continuous trench bordered the North American Continent during the Mesozoic and note that gaps in the fossilized remnants of this trench occur precisely at the location of present-day transform margins (Fig-27).

Accounting for the noticeable bend in the terranes of the Prince William Sound area in terms of their transport histories is a bit more difficult. Geologists refer to

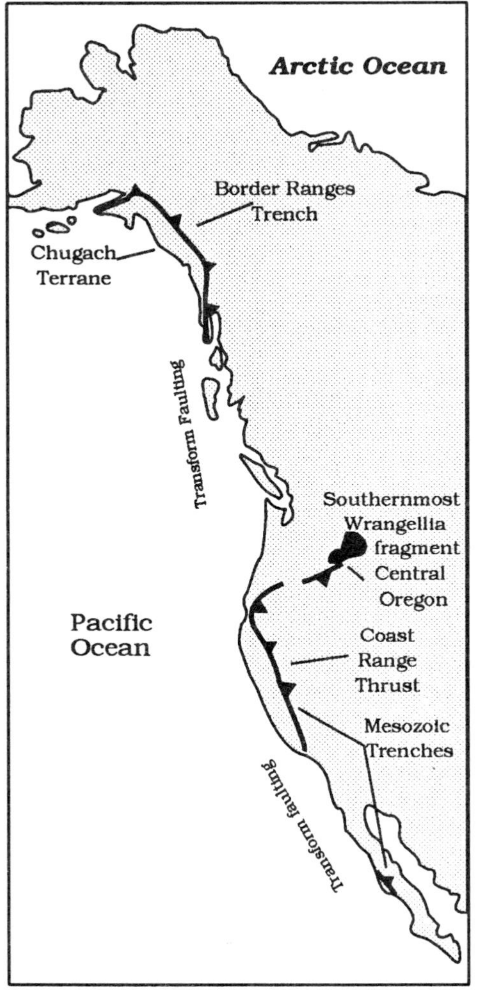

Fig-27. The Border Ranges trench may represent a Mesozoic trench displaced northward by strike-slip faulting. Adapted from Jones, et.al. (1978).

this bend as the *"Alaskan orocline."* Some geologists attribute this bend in Alaska to the convergence of the of the Eurasian and North American plates west of the Bering sea during late Cretaceous and early Tertiary times (Thrupp and Coe, 1986). A second possibility is that the Kula Plate on which these terranes rode is thought to have undergone significant counter-clockwise rotation between 55 and 56 million years ago; and plate motions are known to have assumed a more westerly direction when the composite Kula Pacific Plate replaced the Kula plate 43 million years ago (Lonsdale, 1988, Wallace and Engebretson, 1984). If these terranes were still coupled to the oceanic plate at this time, this mechanism might explain their being bent around the orocline. A third possibility is to note that the Prince William Sound area at the apex of the orocline occupies a unique position at a perpendicular transition zone between strike-slip motion along the Queen Charlotte-Fairweather transform fault system and plate convergence at the Aleutian Trench. Abrupt cessation of this strike-slip motion at the trench may lead to stresses which could rotate crustal blocks. Great arcing strike-slip faults such as the Denali and Castle Mountain-Bruin Bay fault systems far inland may also be due to this interaction. Such stresses would be especially great when a terrane strike-slipping up the coastline encounters the trench as is happening presently with the subduction of the Yakutat Terrane. One can easily imagine how terranes arriving in such a manner and experiencing such stresses might be rotated counterclockwise, further telescoped and bent into a broad, thin arc as they appear today. Some geologists who assume a static geological history for the region explain the arcing shapes of these terranes as being merely the product of arc magmatism such as that occuring in this region today.

The oldest terrane of the Southern Alaska region is the *Alexander Terrane*. The Alexander Terrane lies mostly in Canada and is presently occupied by the St. Elias Range of Mountains. It abuts Wrangellia on its western end, is bounded on the north by the Denali Fault and on the south by the Border Ranges Fault. It arcs from here in a southeasterly direction accounting for a majority of the islands and passages of Southeast Alaska. The basement rocks of this terrane are over 500 million years old and were probably part of an ancient island arc. Its episodic limestone and volcanic beds suggest a long history as an island arc with repeated episodes of uplift and subsidence. By 309 million years ago it was amalgamated with the Wrangellia Terrane to form the precursor of the Southern Alaska Superterrane (Gardner. et al. 1988) (cf. Fig-22).

Wrangellia's rocks, dating back to Carboniferous times (over 300 million years ago), tell a story of island arc volcanism in an equatorial region. During the Triassic, Wrangellia experienced outpourings of great quantities of basalt suggesting, like Iceland today, it may have been for part of its history located over an oceanic spreading center; or like Oregon 17 million ago, it rifted apart to be inundated by flood basalts. About 140 million years ago during the mid-Jurassic, the combined Alexander/Wrangellia Terrane collided and amalgamated with still another terrane, the Peninsular Terrane of Southwestern Alaska, forming the *Southern Alaska Superterrane*.

The Peninsular Terrane consists of the Alaska Peninsula, thin slices of the northwest shores of Kodiak and Afognak Islands, the Cook Inlet area and the western and central Talkeetna Mountains where it stabs deeply into Wrangellia. It is bounded on the northwest and west by the Denali and Castle Mountain-Lake Clark Fault system and on the east and southeast by the Border Ranges Fault system (Fig. 22). Its rocks date from about 240 million years ago and tell the story of its growth as an island arc in the equatorial Panthalassian Ocean 210 million years ago. Its earlier geologic history seems to be distinct from both that of Wrangellia and the Alexander Terrane.

Just as the early histories of these terranes differ before amalgamation, so do their later stories when the Superterrane collided with North America at an oblique angle probably near present day southeastern Oregon or northern California around 100 million years ago. Carried aboard the ancient Farallon Plate, Wrangellia and the Alexander Terrane most likely formed the leading edge. A late Jurassic-Early Cretaceous flysch terrane intruded by andesitic volcanism known as the *"Gravina-Nutzotin Terrane"* appears to have formed on the leading edge of the approaching Superterrane. This small terrane is now drawn out into two elongated segments — the northern most fragment attached to the northeastern edge of the Alexander Terrane at the Canadian border (Fig-22) and the southern fragment adhering to the eastern edge of the Alexander Terrane in Southeast Alaska. This terrane contains conglomerates and flysch that seem to be derived both from the Superterrane and the North American Continental margin at the time of collision (Nokelberg et. al. 1985). A spreading zone developed in the Farallon Plate about 85 million years ago (Wallace and Engebretson, 1984) allowing the Superterrane to move northward with the newly created Kula Plate (Cf. frontpiece plate maps). A small fragment was torn from Wrangellia and adhered to the continental plate in the Hells Canyon area; the rest of the Superterrane remained with the ocean plate which slowly lurched its way abrasively northward. As the two plates slowly ground past one another, a second collision occurred tearing off fragments of the Superterrane and depositing pieces of Wrangellia now known as Vancouver and the Queen Charlotte islands. The major portion of Wrangellia lurched rapidly northward with the Superterrane until coming to comparative rest at its present location in the Wrangell mountains of Southern Alaska. As Wrangellia scraped its way northeasterly against the North American continental margin, the compressive forces of oblique collision combined with northward driving forces of the ocean plate caused both the rocks within the terrane and those of the continental margin to fracture forming elongated, north-south trending faults parallel to the continent. Great slivers of crust were then displaced northward along these faults.

The Peninsular Terrane riding outboard of the Superterrane was most probably deformed differently. The jolting collisions of inboard Wrangellia with the continental margin must have translated the western half of the megaterrane northward along strike-slip faults. Most likely the northern and western edge of the Superterrane collided with the proto-Alaskan margin head on at a subduction trench (the paleo-Aleutian trench). The collision probably closed a former Jurassic-Cretaceous ocean basin which was squeezed between the approaching megaterrane and the former Alaska continental margin. The collision effectively jammed the old Aleutian trench forcing it to "jump" to near its present location. The terrane were later bent in an arc around the orocline.

Outboard of the Superterrane lies an elongated arc of mainly sedimentary and weakly metamorphosed sedimentary rocks (flysch) with minor basalt intrusions, known as *"the Chugach Terrane."* The sedimentary rocks of Chugach Terrane

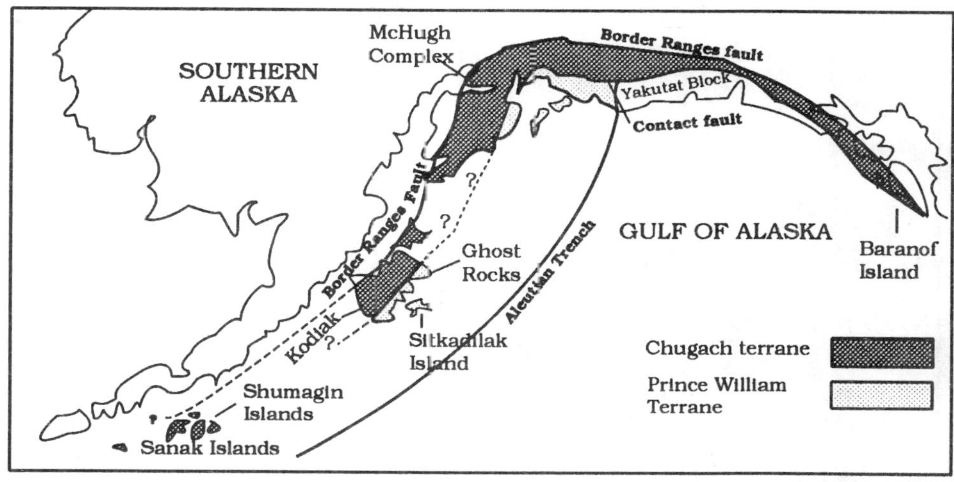

Fig-28. The Chugach/Prince William Terrane.

appear to have been accreted as turbidites in a trench environment between 80 and 65 million years ago and stretch 1200 miles from Baranof Island, across the Chugach Mountains, Prince William Sound, the Kenai Peninsula, and Kodiak Island to Sanak Island to the southwest (Fig-28). The sediments forming these marine rocks may be derived from eroded, volcanic rocks of Wrangellia and the Peninsular Terrane (Dumoulin, 1987). That Wrangellia and the Peninsula terranes might be the source terrane for the Chugach Terrane is also suggested by rocks discovered in the McHugh trench Melange (Nelson et. al. 1985).

Paleomagnetic studies of Chugach rocks indicate that since their creation, they have probably migrated at least 1200 miles north (Grommé et. al. 1981). The Chugach sedimentary rocks are separated from the Superterrane by an older melange belt (the Kelp Bay, McHugh, Seldovia, and Uyak complexes). This melange belt, bordering the sedimentary terrane, seems to represent parts of an ancient subduction trench where the submarine canyon-slope and inner-fan turbidites of the Chugach sedimentary terrane began to accrete as a sedimentary prism onto Wrangellia and the Peninsular Terrane during the late Cretaceous (about 80 million years ago). A reverse fault known as the Border Ranges Fault has developed along the zone of weakness in the Earth's crust created by the former subduction trench. Blueschists along this fault dating from the early Jurassic (200 million years ago) suggest it may mark the site of an even older trench associated

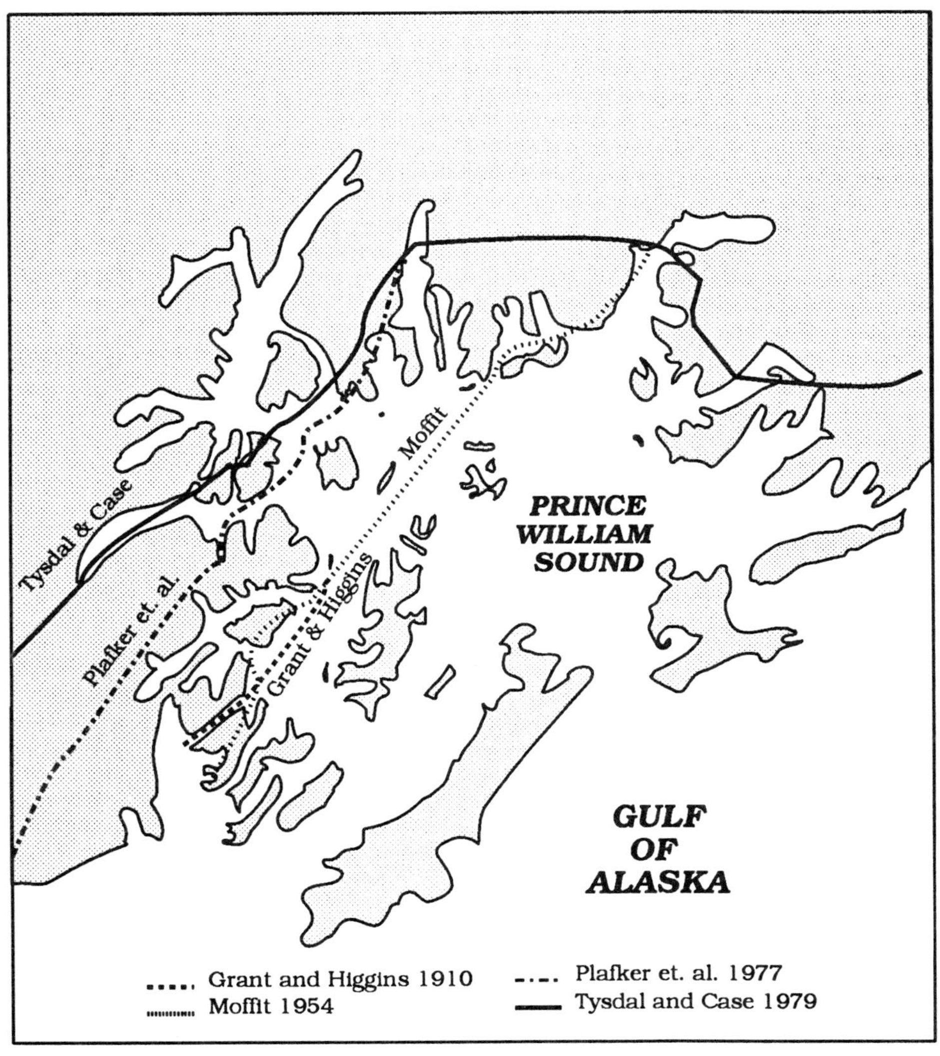

Fig-29. Various suggested locations of the Contact Fault in the western Sound. After Dumoulin (1987).

with subduction along the Superterrane (Roeske et. al. 1989) before its collision with North America. This fault divides the Chugach melange and Chugach sedimentary terranes from the basement rocks of the Superterrane. When one drives the Glenn Highway from Anchorage to Valdez along the Matanuska River, he is tracing out the course of the Border Ranges Fault.

The fault boundary defining the southern extent of the Chugach Terrane is not so easily identified, however. Traditionally, geologists have attempted to distinguish between two groups of rocks in Prince William Sound — the Valdez and Orca groups. The older Valdez group consists of the rocks of the Valdez area, the Port Wells and Passage Canal areas, and the central Kenai and Chugach ranges. The remaining shorelines and Islands of Prince William Sound were referred to as the "Orca Group." Former geologists identified a prominent fault system in the eastern Chugach Mountains and on the northern shores of Prince William Sound which they believed marked the dividing line between these two similar rock groups. They called this fault the *"Contact Fault."* Although the Contact Fault System is easy to trace in the eastern and northern Sound, it seems to disappear in the western Sound. In fact there is little agreement of where it should be placed (Fig-29).

When geologists in the late sixties discovered Paleocene to Eocene fossils in Orca rocks just north of Galena Bay, they concluded that the Contact Fault represented a terrane boundary dividing the Cretaceous Chugach Terrane from a younger Tertiary *"Prince William"* terrane. They included in the Prince William Terrane the Ghost Rocks Formation on the eastern shores of Kodiak Island and formations on Sitkalidak Island (Fig-28). Many of the studies of the Prince William Terrane have been done on Kodiak Island and it is assumed, here, that these results apply also to the Orca Group rocks of Prince William Sound. The Ghost Rocks Formation seems to correlate with the interlayered basalts and turbidites of the Sound while the sandstones of the Sitkadilak formation appear to correlate with the younger, Eocene sandstones of the Montague Belt (Dumoulin, 1987). The axis of the folds of the rocks of these latter two groups tend to point uncharacteristically landward rather than seaward as one would expect for rocks normally formed in a subduction zone.

It was only natural that when terrane theory came along that geologists would want to preserve this historical distinction between the Orca and Valdez groups of rocks by identifying two distinct terranes separated by a fault boundary. However, the similarity on a regional scale of these these two rock groups, which both appear to be weakly metamorphosed trench-fill turbidites, and the lack of a definite suture zone in western Prince William Sound have led other geologists to question the notion of two distinct terranes. (Nelson et. al. 1985, Dumoulin, 1987).

The distinction between the two groups was based partly on a belief that they represented rocks of two different ages and partly that they represent two distinct mineral provinces with gold occurring mainly in the Valdez group and copper and related minerals in the younger Orca rocks. It was also believed that rocks of the

Valdez group are more highly metamorphosed than those of the Orca group. The age distinction comes from a rather sparse and inconclusive fossil evidence which far from establishing a distinctive, sharp break in ages of rocks on either side of the Contact Fault seems only to suggest that the rocks become progressively younger to the south as would be natural for rocks formed through steady accretion (Steve Nelson oral communication). Since both groups of rocks consist of repetitive sequences of trench-fill turbidites, they are difficult to distinguish in the field. It is not clear presently that a uniformly higher degree of metamorphism does exists in the Valdez group. Recent mineral studies seem to indicate that the two groups are not as mineralogically distinct as was once thought. A recent study comparing sandstone compositions across the assumed locations of the Contact Fault in Prince William Sound indicates that there is little to distinguish rocks on either side of the proposed suture zone (Dumoulin, 1987). Finally, like the rocks of the Chugach Terrane, those of the Prince William Terrane indicate formation at a latitude some 20°- 25° south of their present location (Plumley et al, 1983). Because the rocks of the two groups seem to fail two of the three tests to qualify as separate terranes (i.e. they do not seem to be entirely fault bounded entities, and they do not seem to be markedly distinct) this study assumes that the rocks of Prince William Sound formed as a single, continuous accretionary event and we will refer to the resulting single terrane as the *"Chugach/Prince William Terrane."*

What does distinguish these two rock groups to some extent is that the younger, middle fan turbidites of this accretionary terrane seem to have been the space and time nexus of a ridge-trench interaction. About 50 to 60 years million years ago, great quantities of molten basalt poured forth from cracks in the ocean floor mixing with the accreting trench sediments giving the middle section of Prince William Sound its distinctive geological character. One should note, however, that in the area of Jacks Bay in eastern Prince William Sound, a region of interlayered basalts and turbidites seems to extend across the Contact Fault for a significant distance into rocks classified as Valdez Group. Also, it is probably noteworthy that the sediment-hosted copper ores of the Midas Mine, which lies along this belt of mafic rocks in the Valdez Group, are similar to those found at Ellamar and on Latouche Island in the Orca Group.

Often, geological scenarios constructed to explain the origin of Prince William Sound's rocks assume that the Chugach and Prince William terranes are distinct and were swept in on a pacific plate to be docked at a subduction zone against the Alaskan Continental margin in two distinct pulses — the first in the late Cretaceous (the Chugach Terrane) and the second in the late Paleocene or Early Eocene (the Prince William Terrane) (e.g. Plafker, 1988). However, if one rejects the two-terrane concept, then a different scenario is required[1].

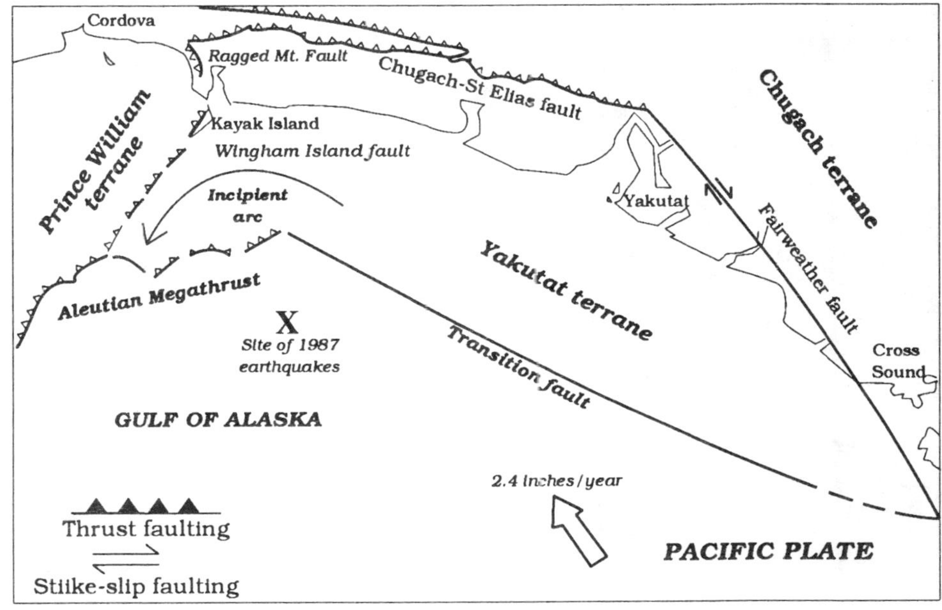

Fig-30. The subducting Yakutat Terrane Adapted from Davis and Pflaker (1986).

Outboard of the Chugach Terrane, just east of Prince William Sound, is the presently accreting Yakutat Terrane. The Yakutat Terrane is currently subducting in the Aleutian trench in the area off Kayak Island. This terrane appears to be a composite terrane consisting of an eastern, Mesozoic flysch and melange subterrane and a western subterrane underlain by Paleocene/Eocene basaltic ocean crust. Geologists believe that the Yakutat Terrane may represent a basaltic chunk of mid-ocean spreading center that plowed into the North American coast some time during the Eocene. During the collision, it tore off a fragment of the Mesozoic coastline and was subsequently translated northward by strike-slip faulting as a composite terrane (Bruns, 1983). The Yakutat Terrane seems to have moved northward along the Queen Charlotte-Fairweather transform fault system — probably from the vicinity of British Columbia. Its position outboard of Chugach/Prince William Terrane indicates a later arrival date. As the movement seems to have been oblique to the continental margin underthrusting it, geologists speculate that as much as 135 miles of the terrane may have already disappeared beneath the Chugach-St. Elias fault system and the Kayak Island subduction zone within the

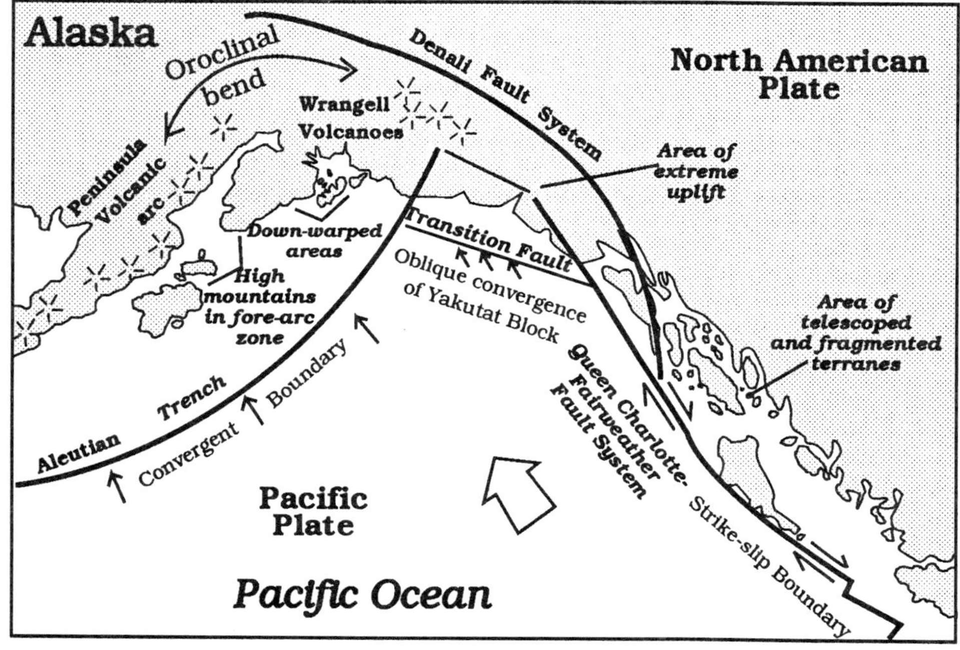

Fig-31. Major geographical features of southern Alaska resulting from present plate tectonics. Adapted from von Heune, et.al. (1978).

last 20 million years (Plafker, 1988). This underthrusting of the Alaskan continental margin has resulted in the spectacular uplift and deformation in the St. Elias and Chugach Mountains. Because uplifted mountain ranges do not ordinarily occur between a subduction trench and its magmatic arc, geologists believe that the subducted portion of the Yakutat block may presently underlie Prince William Sound, the Kenai Mountains and Kodiak Island and accounts for their uplift. The shallow angle of subduction in the Prince William Sound area may also be due to the presence of this accreting terrane (von Huene et. al. 1985)

Currently, the Yakutat Terrane is partially coupled to the Pacific Plate and moving northward with it at a little over 2 inches/ year. Its western edge is thrust against the Chugach/Prince William Terrane undoubtedly explaining some of the folding in the rocks of eastern Prince William Sound that seems to have been occurring for the last 15 million years (Helwig and Emmet, 1981). The compressive forces created by the convergence and subduction of the Yakutat block at the

Aleutian trench also seem to be temporarily downwarping the Prince William Sound and Kenai Fiords areas. Here, we find numerous drowned glacier valleys and cirques creating a shoreline different from both the elongated fiords of the strike-slip zone of Southeast Alaska and the underthrust zone of the Yakutat area.

In the Yakutat Terrane we may be witnessing in miniature what may have occurred as other Southern Alaskan terranes reached this transition zone between strike-slip movement and subduction million upon millions of years ago. Like them, the Yakutat Terrane seems to have been telescoped and elongated in a northwest-southeasterly direction. And like them the terrane has, on reaching the area of the orocline, undergone counterclockwise rotation — in this case about 20° (Plafker, 1988). At its western end, where it is thrusting against the Chugach/Prince William Terrane and subducting in the Aleutian trench, a bend appears to be developing (cf. Fig-30). One wonders whether tens of millions of years in the future if the Yakutat Terrane will appear smeared out in a thin arc around the oroclinal bend much as are the Peninsula/Wrangellia and Chugach/Prince William terranes today.

Periodic large earthquakes in the vicinity of Yakutat offer violent testimony of the strike-slip movement and convergence of this terrane with the Southern Alaska continental margin. The area of the Transition Fault off Icy Bay and Cape Yakataga has recently been of great interest to geologists because it has been seismically relatively quiet since 1899 when a great earthquake shook the region. A long period of seismic quiescence may be a sign that large stresses are building up in this section of the fault. The great earthquake of 1899 was preceded by a number of severe shocks. Near the end of 1987, we experienced in a relatively short period shocks of 6.9 and 7.5 on the Richter scale suggesting that the Prince William Sound region might soon expect another large quake.

Chapter 2. Geological History of Prince William Sound and Southern Alaska

What follows might be called "geo-fiction;" however, like most good fiction it attempts to mirror reality as faithfully as possible. Geologists armed with new concepts, new exploration techniques and new data have only recently begun to unravel the complex and confusing geological history of Prince William Sound and Southern Alaska. Many details of the following narrative will undoubtedly change in the next few years. In order, not to interrupt the narrative, I have not included in the text lengthy discussions of sources and debatable points but have footnoted this material for my more technically minded reader.[2]

Carboniferous and Permian Beginnings

It was a warmer time than ours. Three hundred and fifty million years ago, the North American Continent lay much closer to the equator, its western half submerged beneath a great shallow sea, its eastern region a vast, swampy lowland. Eastern North America resembled more the Congo Basin than the present rolling hills of Appalachia. Here, the sweltering tropical humidity had given birth to a riot of strange vegetation. Fantastic trees covered with a fish-scale bark clung desperately to the richly rotting humus — their pulpy stalks struggling hundreds of feet in the air, gasping for light in an attempt to penetrate the dense forest canopy above. Great tree-ferns overtopped these. In the damp rot of this flowerless forest, evolution favored the tall. Lower down in the gloom, crept giant clumps of club mosses forty times the size of current species while slender horsetails, over eighty feet tall, stretched up toward the tenuous light.

Still other exaggerated species inhabited the vegetative excess of the Carboniferous coal forests. Gargantuan, two-foot long dragon flies flitted bird-like through the murky realms while cockroaches the size of mice stole through the dense undergrowth. These insects fed on the abundant plant life while giant scorpions and spiders fed on them. But even these were not exempt from the food chain for only recently some obscure fossil fish had climbed upon the land, front fins forward, gulped its first breath of atmosphere and become the first amphibian. These ponderous reptilian prototypes were not very graceful or efficient as they slid sled-like through the slime-filled swamps, but fossil remains of their ten-foot long, five-hundred pound bodies indicate that they thrived quite nicely in their new, terrestrial environment. In fact, they were millions of years later to evolve into one of the most successful species ever to inhabit the planet — the dinosaurs.

During the later Carboniferous and adjacent Permian, the Earth's continental masses were drifting together on a colossal collision course. Eurasia drifted eastward closing an ocean basin to slam into North America. This fused mass then rotated into a collision with Gondwanaland coming up fast from the south. By 250 million years ago, the megacontinent, Pangaea, had already formed. This great rearrangement of the Earth's crust apparently led episodically to chilling climatic effects as witnessed for example in the great Permian ice sheet that covered most of southern Gondwanaland stretching to within perhaps 30° of the ancient equator.

Plate tectonics tells us that plate motions are global phenomena. Plate movements on one part of the globe imply motion elsewhere. What is created at a rifting zone must be destroyed in a subduction zone. The same plate motions which were bringing the continents together were preparing the birth of Southern Alaska somewhere far away near the equator in the eastern Panthalassian Ocean.

The Birth and Development of Wrangellia (325-245 Mya)

Sometime in the Carboniferous, a little over 300 million years ago in a period geologists refer to as the "Pennsylvanian," for it was then when the great coal beds of Appalachia were deposited, a mighty tectonic battle was being waged near the Panthalassian equator. The adversaries, however, were not two continents but two ocean plates moving slowly but inexorably in opposite directions. The outcome of the conflict was already predetermined by simple physics — the older, colder, denser plate must thrust beneath the younger, warmer, lighter plate. A profound ocean trench marked the zone of collision. For millions upon millions of years, the underthrust plate inched down toward the mantle carrying its load of deep ocean sediments.

On reaching a depth of about 60 miles, partial melting of the sediments and peridotite of the upper mantle began to occur. The melt began fractionating off lighter minerals containing aluminum and silica. This less dense melt, having the composition of andesite, then slowly began to rise, wedging its way up through the lithosphere. Finally, with a great hissing roar, the magma plume pierced the surface extruding white hot andesite onto the ancient ocean floor. Thus the first rocks of an ancient island arc which was later to become part of Southern Alaska were born.

The growth of Wrangellia was a ponderously slow process. For millions upon millions of years, molten lava poured out through vents bordering the subduction zone, encountered the cold waters of the ocean abyss and hissed itself into solid rock. Gradually, eruption by eruption, a hundred miles back from the trench on the overthrust plate, a gigantic, underwater, volcanic massif arose. Paralleling the trench, this elongated volcanic arc slowly reached for the ocean's surface miles above. Occasionally, the whole region would tremble from earthquakes generated deep down in the subduction zone; or more often, shallow tremors would shake the volcanic roots as molten rock rumbled its way up from the bowels of the earth. At these times, massive accumulations of ash and unconsolidated sediments would avalanche down the steep slopes, slice across the trench, and spread out over the abyssal plains. From time to time, granitic plutons would intrude the great volcanic arc further uplifting it and adding to its mass.

One of these plutons that intruded the island arc 309 million years ago, we find today just west of the Yukon-Alaska border on the flanks of Bernard Glacier. This Pluton intrudes both the basement rocks of Wrangellia and the basement rocks of a much older island arc terrane — the Paleozoic Alexander Terrane of the Yukon Territory and Southeast Alaska. The stitching together of these two terranes by this pluton suggests that the Wrangellian island arc formed on the basement of the Alexander Terrane or that it collided with it at an ocean trench early in its geological history (Gardner et al. 1988).

Sometime in the Late Pennsylvanian or early Permian, parts of the Wrangellian subterrane at last penetrated the ocean's surface. An arc of new tropical Islands was born. But the new islands' triumph were not to last. What was born in a moment of spectacular volcanic violence would be destroyed by eons of almost imperceptible but relentless erosion. While rainstorms, drop by drop, washed away at the soft volcanic rocks on the mountain slopes, the ocean surge tore away at the shoreline. The magma supply diminished; cauldras collapsed; and the arc subsided beneath the restless waves.

As the islands sank, wave action leveled off their tops so that all that remained of the former island arc was a great submarine plateau. As the plateau steadily sank in Permian times (about 290 mya), various ocean creatures with calcium carbonate shells and skeletons began to inhabit the shallow depths initiating a process which would lead to a slow upward growth of carbonate reefs. These skeletal materials accumulating around and atop the plateau were the raw material which would later be compressed into the Permian limestones found in the Wrangell Mountains today. There were many larger species in a position to donate their calcacerous frames to this endeavor — various forms of clams, sea snails and the large, coiled jet-propelled ammonites. But the majority of the enterprise was left to a small, microscopic plankton known as "*foraminifera*." Billions upon billions of their tiny globular or spindle-shaped skeletons rained down upon the submarine plateau. Occasionally, this steady, even deposition of sediments would be punctuated by submarine eruptions which would scatter a fine, volcanic debris over the carbonate mass leaving a thick bed of Permian limestone intercalated with volcanic sediments.

For tens of millions of years, these sediments accumulated as the arc struggled between subsidence and growth. Then, suddenly at the end of Permian times (245 mya) deposition ceased. A great environmental catastrophe had struck. Whether the Earth's orbit had encountered a large comet like Halley's progressing wrong headedly around the Sun, or a giant asteroid had struck the Earth, or merely the great continental collisions had drastically altered the climate, ninety-six percent of all marine life on Earth was obliterated.

Life on land fared little better. The great cooling off and drying out of the land that differentiated the Permian from the warmer, moister Carboniferous created extensive, red, sandy deserts replete with improved life forms to inhabit them. The pulpy scale trees and giant tree ferns had given way to hardy conifers. A strange breed of amphibians had invented a new hard-shelled egg which sealed in moisture allowing its bearer to escape the tyranny of the swamps and evolve into the first land dwelling reptiles. Certain of these versatile and forward looking types would later develop mammal-like characteristics. Many but not all of these perished during the great dying out at the end of the Permian.

The Triassic period (245-208 mya) which followed marked the beginning of an era of great evolutionary inventiveness. The great Permian catastrophe had transformed former ecological niches into ecological chasms. Those species that survived spread rapidly and prolifically while exhibiting a good deal of evolutionary imagination. Onshore, reptilian life exploded into numerous and variegated species. It was at this time that the first dinosaurs emerged. There was a proliferation of conifers; and palms developed. In the oceans, clams, oysters and coiled ammonites prospered and grew in size. Many forms of coral thrived; and there was a resurgence in many limestone building species.

The Triassic was especially significant for the Wrangellia portion of the Wrangellia-Alexander composite terrane; for it was during this period that the Wrangellia would acquire much of its mass and would once more break the ocean's surface. However, it was neither the uplift of submarine andesitic volcanism nor the steady accumulation of shallow, marine sediments that would accomplish this, but a wholly new event. It started with a swarm of shallow quakes, then the sea floor beneath the arc began to stretch, thin and finally tear apart. The volcanic island arc had found itself situated over a rifting zone at a latitude of 15°.

For five million years, there was a vast extrusion of molten basalt from the rifts in the sea floor splitting the arc. At first, the basalt poured forth quietly but voluminously in familiar elongated pillows or crystallized slowly as gabbro plutons within the former volcanic pile. So great was the outpouring that the mass soon lifted above the sea's surface no longer forming pillow basalts but extruding instead in a distinctive subaerial form. In extent, the extrusions of these Nikolai greenstones (basalts) rival those of the later Columbia River plateau and are in places up to three and a half miles thick. Wrangellia had once again achieved island status. However, when the great upwelling from the mantle ceased, the island mass began to subside. Once again over countless eons, the great island sank under its own weight and the surrounding sea invaded its shores. Once more tiny marine organisms spread their carbonate skeletons over the submerged mass to provide material for future limestone beds which would be scattered from the Idaho-Oregon border, to Vancouver Island to the Wrangell Mountains of Southern Alaska.

While volcanism had ceased in the Wrangellia portion of the Wrangellia-Alexander Terrane by mid-Triassic times, volcanism continued in the Alexander Terrane throughout the late Triassic into the Jurassic.

The Peninsular Terrane / the Formation of the Superterrane (245 - 144 mya)

During the Triassic another extensive island arc inhabited the equatorial regions of the eastern Panthalassian ocean. A comparison of Triassic fossil shells from this terrane with those of Wrangellia suggests that the two might not have been far apart (Newton, 1983). The history of island arc volcanism in the Peninsular Terrane was in many ways similar to that of Wrangellia, however the copious outpouring of Triassic basalts is noticeably absent. Evidence from blueschists (cf. p. 28) in the Peninsular Terrane on Kodiak Island, off Seldovia in the Kenai Peninsula and along the Border Ranges fault near Tazlina lake suggest that the subduction zone which formed this island arc was still active 190 million years ago at the beginning of the Jurassic.

The Jurassic period is divided from the Triassic by another great catastrophe to life on earth — the so-called "Norian event." Although the Norian event decimated both marine and terrestrial species, it failed to rival the earlier Permian catastrophe. A suspicious asteroid crater in the Canadian province of Quebec dates from this time and is our only clue as to a possible cause.

Once again, however, life on Earth rebounded with a vengeance. During

Jurassic times, the marine realm, dominated by ammonites, saw a resurgence of all manner of clams, snails, corals, sea urchins and stalked sea lilies. New forms such as modern lobsters, crabs, crayfish, sharks and skates emerged; while strange, Loch Nessian reptiles reenvaded the the sea realm to reenact their piscine pasts.

Ashore, great Cycad forests spread over the landscape. A Cycad tree to a modern botanist would appear to be some kind of strange chimera with its pineapple trunk, its fantastic fronds that unrolled like those of a fern, its primitive flowers and its conifer-like cone at the top. More familiar ferns and various conifers, however, also inhabited these bizarre forests.

But the real Jurassic innovators were the dinosaurs. While they shared the swamps with prolific numbers of primitive crocodiles and turtles, they became totally dominant elsewhere. Whatever the challenges of the environment, their genes responded to them. They inhabited the seas, the swamps, the forests, and the deserts. Even the aerial realm was no stranger to them as they soared through it like strange, leathery gliders. No size seemed too large or too small for them. If sharp teeth were required to tear the flesh of their prey, they developed them; if a leathery hide was necessary for protection, they grew one; if a bipedal stance was necessary for fleet pursuit, they assumed one. They developed adaptations for feeding on the slime and the roots of the swamps, on the mollusks of the seas, on the insects of the deserts and on the highest branches and leaves of the forests and, last but not least, on each other. Their evolutionary instinct was phenomenal for they developed strategies that would outwit the great catastrophe at the end of the Cretaceous (65 mya). By transforming their primitive scales into precursors of fur and feather and most importantly inventing warm bloodedness, the dinosaurs were able to project their genes into the future. It was in the Jurassic that a tiny, mouse-like dinosaur, first dropped to the forest floor to become a creature of the night and take the first steps toward mammaldom.

During the Jurassic there was a great stirring in the land. The megacontinent, Pangaea, which had only recently attained final assemblage began to show signs of instability. Great, ragged rifts began to rend the Earth's surface tracing out the boundaries of the present continents. Where rifts appeared, basaltic magma poured forth inundating the land. Pangaea was literally being torn apart. Apparently, the great continent had lasted long enough. It was time for the restless plates to get on with their task of randomly rearranging the Earth's crust.

Such was the state of the world as volcano after volcano erupted along the Jurassic arc of the Peninsular Terrane. Inch by inch, the ocean floor thrust beneath the ancient trench, fueling the volcanoes of the arc. There was a period of prolific volcanism all along the arc 200 to 170 mya. However, between 176 and 154 mya much of the lava generated in this subduction zone never saw the light of day but

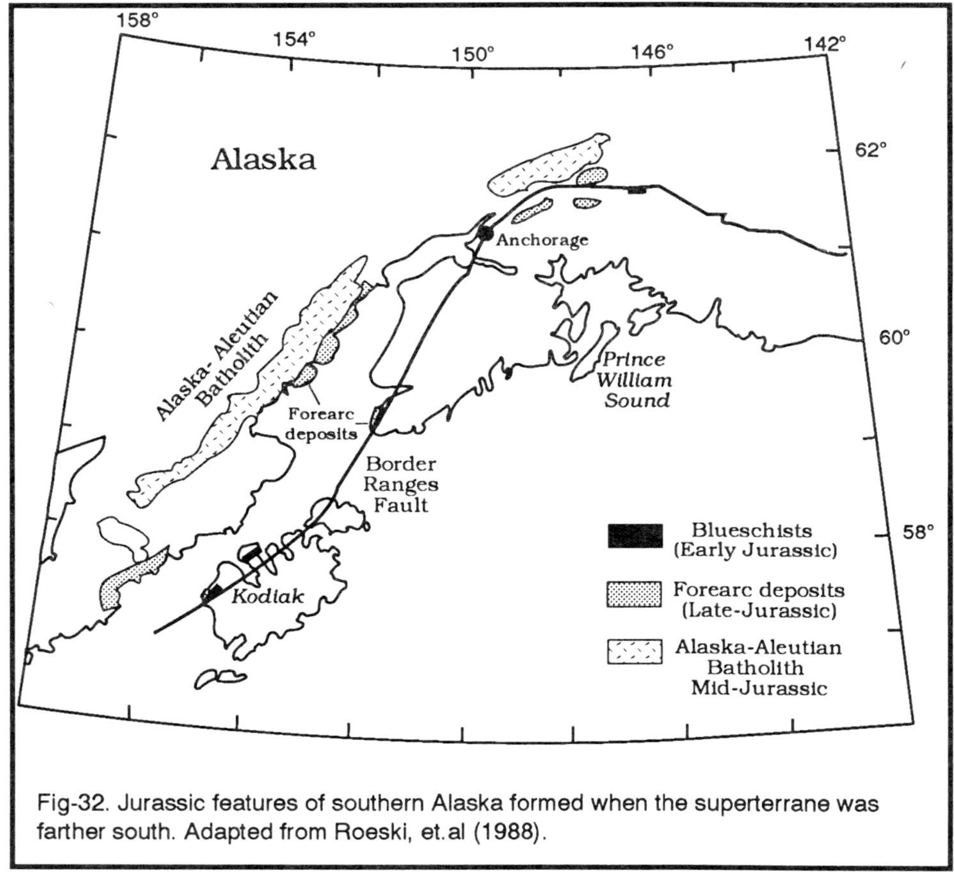

Fig-32. Jurassic features of southern Alaska formed when the superterrane was farther south. Adapted from Roeski, et.al (1988).

cooled slowly beneath the surface as a great, granite batholith. These plutons, formed so many millions of years ago in the southern ocean, have been subsequently unroofed and now appear far to the north as the Alaska-Aleutian Batholith. Sedimentary rocks formed from the eroded sediments of these island arc volcanoes, have also been identified across from Kodiak Island, on the west shores of Cook Inlet, and in the Matanuska Valley. The shift from volcanism to plutonism suggests that the crust of the arc must have reached such a great thickness that the magma failed to penetrate the surface. We must imagine the Peninsular Terrane as an arc of lofty mountains presiding over the ancient southern ocean.

For millions upon millions of years, as the dinosaurs began taking possession of the dispersing continents, the plate carrying the Peninsular Terrane moved steadily in a northerly direction. As the plate advanced northward, the subduction zone on the leading edge of plate (the same Jurassic subduction zone which had formed the Alaska-Aleutian batholith of the Peninsular Terrane) slowly but steadily consumed the older basaltic sea floor thrust into it by the southward moving plate. This plate was propelled by a spreading center far to the north. Rafted on the south-moving plate rode a terrane which would later be known as "Wrangellia."

Gradually, gradually at the rate a fingernail grows, the two island arcs, each riding on its respective plate, approached the trench as more ocean floor was consumed beneath the Peninsular arc. Finally, the subduction zone became the site of a ponderously slow but violent collision. The Peninsular arc astride the younger, lighter plate was thrust northward up and over southward-moving Wrangellia. The subduction trench was closed, and Wrangellia became sutured to the Peninsular Terrane. Suturing at the trench resulted in a rearrangement of plate boundaries, and the new Superterrane moved off in a northerly direction. The collision of the two terranes most likely resulted in an easterly component to the northward motion for within 50 million years, this new micro-continent would slam obliquely into the continental margin of westerly moving North America. With the suturing of the subduction zone, the volcanoes of the Peninsular arc snuffed out (154 mya). As the newly amalgamated terrane moved northeasterly, a new subduction zone opened on its easterly, leading edge and renewed Island arc volcanism established itself in the Wrangellia portion of the terrane (Nokleberg et al. 1985). [3]

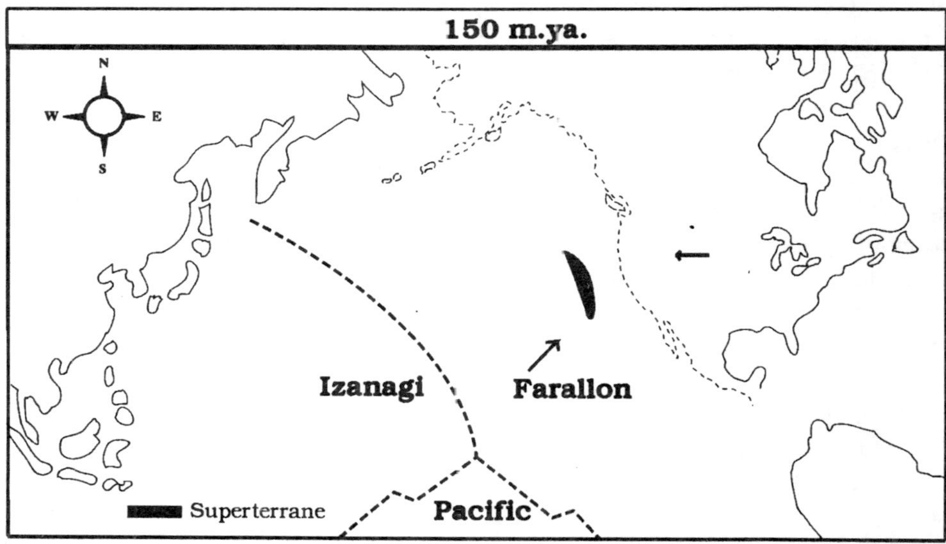

Fig-33. 150 my. Adapted from Wallace & Engebretson.

Thus, most of present-day Southern Alaska and part of Southeastern Alaska moved as a single unit northeastward toward North America. The present relationship of these three terranes suggests that Wrangellia formed the northern and eastern, leading edge, the Alexander Terrane formed the southeastern corner while the Peninsular Terrane formed the the northwesterly, leading edge and the entire western shore. However, as the Superterrane has been severely fragmented, distorted and rotated during its collision with the continent, no one knows for sure its actual Jurassic configuration.

During the Late Jurassic and Cretaceous (150-120 mya), the micro-continent drifted on the ancient Farallon Plate in a northeasterly direction, headed for a rendezvous with North America. Meanwhile, an ocean-filled rift, probably no wider than the present Red Sea, had begun to develop between the East Coast of North America and the west coast of Eurasia. The newly created North American plate began moving in a westerly direction toward the Superterrane.

During the slow trip north, sporadic volcanism shook the Wrangellia portion of the terrane. With the shutdown of the Jurassic trench on the Peninsular, trailing edge of the newly formed terrane, the focus of volcanism and plutonism shifted to its northern and eastern, leading edge. A new trench slanted down under the new leading edge as the great terrane drifted in a northeasterly direction. Turbidity currents from Wrangellia and the Alexander Terrane swept down over the trench. Slowly, accretionary wedges developed and the superterrane scrapped these sediments off the down going plate. Thus, a new accretionary terrane known as the Gravina-Nutzotin terrane developed on the leading edge of the advancing superterrane. (Nokleberg et. al. 1985).

Meanwhile the Peninsular portion, whose subduction trench had been closed by the collision with Wrangellia, remained remarkably quiet. It had become a passive margin. Here, the tropical winds and rains attacked the lofty summits. Finally, after millions upon millions of years, the once imposing Peninsular Volcanoes were being relentlessly reduced to their plutonic roots — their once magnificent slopes now a thick apron of insignificant, volcanic sands skirting the trailing edge of the arc. Nor did the persistent forces of wind, rain, and surf cease once the more resistant granites had been exposed. Even these, they attacked with great resolve, gradually wearing them down also into a characteristic feldspar and quartz-rich sand. Sandstones of this depositional episodes can still be seen along the Alaska Peninsula today (Moore and Connely, 1977).

Evidence that the Superterrane was approaching the North American continent during Late Jurassic to Cretaceous times is preserved in the flysch of the Gravina-Nutzotin Terrane which seems to have been formed from sediments shed *both* from the Superterrane and from North America as continentally derived cobbles appear in some of its conglomerates (Nokleberg et al. 1985). Jurassic/Cretaceous volcanism and plutonism in the accretionary wedge are thought to have been the results of the closing of the Superterrane with the North American Continent. The final collision of the microcontinent with North America most likely occurred sometime in the mid-Cretaceous around 110 mya[4] at a Mesozoic continental margin then located along what is presently the Oregon-Idaho border considerably farther inland than the present margin. The terrane probably fused to the continent somewhere near present day Cape Mendocino (Coe et. al. 1985) at a paleolatitude of about 32° (Panuska, 1985). [5]

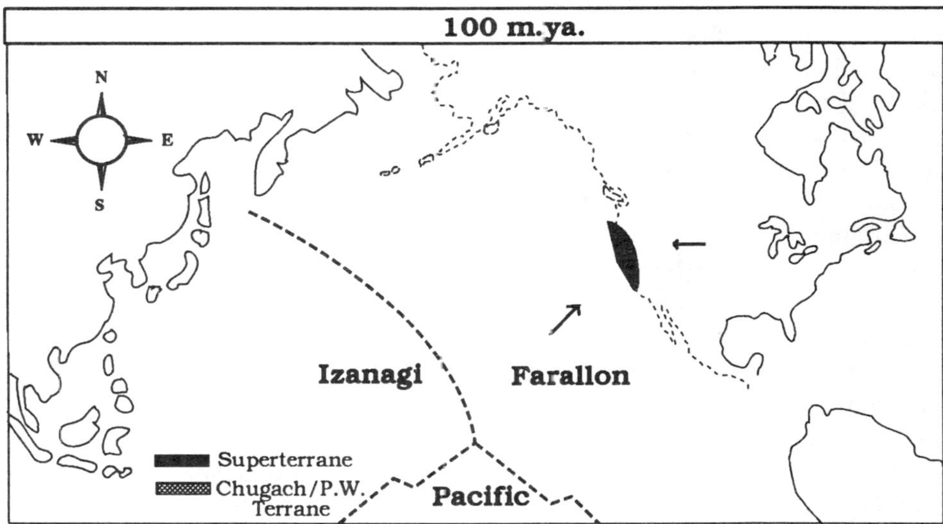

Fig-34. 100 mya. Adapted from Wallace & Engebretson.

Cretaceous Collision and the Birth of the Chugach/Prince William Terrane near Latitude 40° North (144 - 66 mya)

Studies of ancient sea levels suggest that the Earth's climate during the Cretaceous period (144-66 mya) was much warmer than either the preceding Jurassic period or the succeeding Cenozoic era (66 mya to the present). Sea levels were at an all time high, suggesting that little or no ice was locked up in the polar caps. In fact, a broad, shallow inland sea covered much of Western North America from the latitude of present day Alaska to Mexico. In and around this great sea, lived the gigantic dinosaurs whose bones grace many a modern museum. This world of lumbering giants must have been a strange and dangerous place for our early ancestors, the mammals — small, furry nocturnal animals the size of a rat. With their adaptation of fur and warm blood, they had carved an ecological niche out of the night; for many species of dinosaurs, like most modern reptiles, were probably rendered inactive by the disappearance of the sun.

During the Cretaceous, a second, warm blooded species had evolved from the dinosaurs — primitive birds. We must imagine these, however, not as the graceful forms that fill our modern sky, but as lumbering, two legged, dinosaurlike creatures, resembling more an ostrich than an eagle. Both warm-blooded species survived in the warmth of the Cretaceous, but neither really prospered. Except for allowing them the run of the night, their adaptations had no great survival value in this reptilian world ruled by sheer size.

In the oceans most Jurassic species continued to thrive and increase in size. On land the seed bearing plants of the Jurassic continued to prosper and were joined by a new evolutionary invention — the modern flowering plants which quickly worked out a symbiotic strategy with the insect order to disperse their kind.

An early Cretaceous map of the coastline of Western North America would have little in common with its modern counterpart. Much of Alaska, British Columbia, Washington, Oregon, and California would be notably absent. These missing fragments had not yet arrived on the conveyor belt of the ancient Kula and Farallon oceanic plates. In fact at this time, only a single great ocean plate, the ancestral Farallon plate, lay off the west coast and was moving in an easterly direction carrying on its back a number of islands, some large enough to qualify as micro-continents. Some of these like the southern Alaskan Superterrane were formed as island arcs near the equator, others originated as islands in the Tethyan sea, still others were crustal fragments ripped from ancient continents by the ceaseless motion of the plates. Submerged beneath the restless waves rode still other crustal blocks, great basaltic seamounts and submarine plateaus — all on a collision course with North America which was being pushed westward by spreading along the Mid-Atlantic ridge. On its western shore, a deep, offshore trench marked the collision zone where the Farallon Plate dived down under the continental plate. Most likely, this subduction zone bordered the entire coast from proto-Alaska to Mexico and perhaps beyond.

We must imagine the approach of the Superterrane to the North American continent at a latitude of about 30° as an imperceptibly slow and thoroughly inglorious affair. A person inhabiting the huge, offshore island would have been no more aware that he was approaching the continent at the rate of several inches a year than the present residents of Los Angeles are aware that they are approaching San Francisco at a comparable rate. Yet a time lapse camera trained for a hundred thousand years on the island from a point on the ancient Idaho coast would have yielded a remarkable filmstrip.

At first, only wisps of black smoke from the Gravina-Nutzotin volcanoes would have been visible over the horizon; then after tens of thousands of years, the peaks of the volcanoes themselves would appear; and finally, after scores of thousands of years, the rugged Wrangellia coastline of the giant island arc would loom into view. The megaterrane would have had more the aspect of an advancing continent than an approaching island. As the terrane drew near, turbidity currents would avalanche down the continental margin spilling out across both the continental trench and into the Gravina-Nutzotin trench which dipped down beneath the approaching terrane.

Gradually, gradually during the Mid-Cretaceous, the gap between the two trenches closed as Wrangellia approached at an oblique angle from the southeast. Unlike the dense, basaltic ocean crust that had proceeded it into the continental trench, the felsic crust of the Superterrane was too light to be subducted; hence, it was buoyed up and shoved across the trench onto the continental margin. Thus, Southern Alaska became fused to North America at the location of Northern California and Oregon but at the latitude of Baja. It would remain here moving with the North American Continent for another 50 million years (Engebretson, Debiche and Cox, 1983 suggest a somewhat similar scenario). Where the Superterrane collided with the continent a great metamorphic/plutonic welt formed. Later, transported northward by strike-slip motion, these rocks would be known as "the Coast Plutonic complex." The closing of the trenches along the continental margin caused a trench jump to the trailing edge of the Superterrane now forming the new continental margin (Pavlis, 1982). This newly formed trench known as *"the Border Ranges Trench"* probably established itself along a zone of weakness created by a former subduction zone then bordering the western, trailing edge of the Superterrane. Early Jurassic blueschists from the former trench have been uplifted along the present Border Ranges Fault from Kodiak Island to Chichagof Island in Southeast Alaska. Similarly, the wrenching collision of the Superterrane with the North American Continent and consequent trench jump seems to have ripped up some of the lower crustal and upper mantle rocks from the Peninsular Terrane's basement thrusting them to the surface. These mafic and ultramafic rocks now scattered along the Border Ranges Fault for 600 miles from Kodiak Island to near Tonsina and are known as the "Tonsina mafic-ultramafic complex." (DeBari and Coleman, 1989). While a part of North America at a latitude of about 40°, the Superterrane would manufacture the rocks of Prince William Sound in the Border Ranges Trench and then resume its northward journey.

For tens of millions of years, the newly formed trench became a dumping ground for rocks cascading down off the docked Superterrane and for rocks carried in on the conveyor belt of the subducting Farallon plate. From the Superterrane came conglomerates, some as old as Wrangellia itself, fragments of Late Triassic limestones from the Peninsular Terrane, chunks of Jurassic granite from the Alaskan-Aleutian Batholith, and blueschists and greenschists from the former Jurassic trench. From the sea came pieces of Triassic, Jurassic, and Cretaceous chert, chunks of sandstone, and blocks of basaltic seafloor. All these were carried down into the infinite anarchy of the subduction zone where they were tumbled and heated into a weakly metamorphosed mass. Uplifted at a later date near Anchorage, Alaska, they would become known as the McHugh Melange and would represent the early beginnings of the Chugach Terrane.

For thousands of years, sediments from the Superterrane would accumulate off river deltas near the edge of the trench shelf. These would incline heavily, momentarily resisting the slope; finally, shaken by a great quake, they would suddenly avalanche, slicing out a deep canyon in the adjacent trench wall to spread out fan-like over the trench floor and and out onto the abyssal plain beyond. The heavier cobbles would be the first to drop from the turbid flow, usually near the bottom of the trench slope and along the inner portions of the fan, followed by less heavy deposits; and finally, a fine silt would spread out over the far reaches of the fan. For millions of years, this scenario would repeat itself, until finally the turbidite fans reached a great thickness. About 80 mya, during the Late Cretaceous, these turbidity deposits, which were being carried back toward the trench by the ocean plate, were so thick and light that they could no longer be totally subducted. They began to pile up against the inner trench wall to be crumpled and compressed into a great accretionary wedge. Rocks formed from this wedge were mostly con-glomerates, sandstones, shales, and dirty sandstones (greywackes). Sediments carried down into the trench were compressed, heated up and scraped off on the underside of the overriding plate. Uplifted later, these would appear as metasandstones, slates and argillites of the Chugach Terrane bordering northern Prince William Sound.

Why the Valdez Group suddenly began to accrete 30 million years after the reestablishment of the Border Ranges Trench (110 mya) is somewhat of a mystery. Perhaps, only by this time had sufficient sediment loads accumulated to begin the accretionary process. Or perhaps the rate at which these sediments were delivered to the trench accounts for the abrupt commencement of accretionary activity as a result of a major plate reorganization that took place 85 mya. At this time, the eastward migrating Farallon plate suddenly split in half as a new spreading center opened in mid-plate establishing a new triple junction between the Pacific, Farallon, and newly created Kula Plates.

This reorganization of plate boundaries apparently resulted in a rapid, northeasterly movement of the Kula plate from 74 to 56 mya (Wallace and Engebretson, 1984). Since these dates correspond to the dates of accretion of the Valdez and Orca rocks, it is tempting to attribute the onset of accretionary wedge formation to this increase in plate velocity.

The increased sediment loads reaching the trench due to increased plate motions was evidenced in a second way. Plutonic activity, which had ceased about 154 mya, suddenly resurged in the Peninsular terrane. This volcanism and plutonism, undoubtedly the result of subduction along the Border Ranges trench, began abruptly 80 mya and lasted for another 20 million years. One can speculate that the

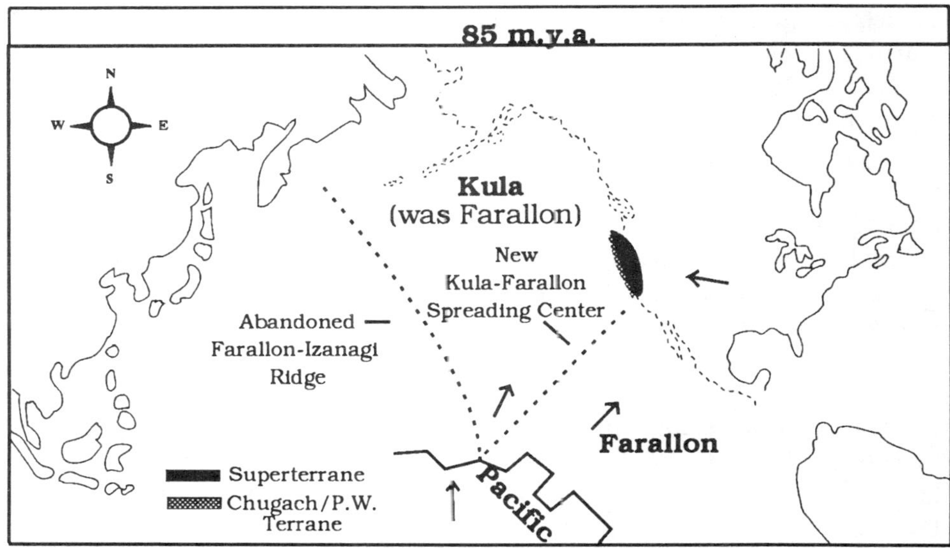

Fig-35. 85 mya. Adapted from Wallace & Engebretson

abundant sediments which accounted for the early, accretionary wedge formation of the Valdez Group, also provided the wet, subducting material necessary for the partial melting which would rekindle the fiery furnaces beneath Peninsular Terrane. Further, the increase in plate motion would stimulate more volcanism and uplift in the Superterrane which in turn through erosion would supply still more sediments for the accretionary terrane, providing still more subducting sediments to feed the volcanism.

The present spatial relationship between the Cretaceous melange belt and the Late Cretaceous and Paleocene plutonic arc appears to be similar to that of the Aleutian Trench and its present magmatic arc. However, we must not imagine that at the close of the Cretaceous the terranes composing the Superterrane or the newly formed melange and flysch belt of the Chugach Terrane had their present latitudes, orientations, shapes, or longitudinal extents. They were later to be fragmented, elongated, slivered, rotated, and bent around the orocline on their long journey north.

The Paleocene Development of the Orca Group: A Ridge-trench Interaction at 40° North (66 - 57 mya).

Slowly, for 15 million years, the rocks of the Valdez Group, the Kodiak Formation, and Chichagof and Baranof Islands accumulated in the Border Ranges Trench — now a part of the ancient subduction zone that bordered the entire North American coast.

Then suddenly, 65 mya, the sky burst, the earth shook, and a great cloud of dark debris shrouded the planet, and the dinosaurs disappeared from the face of the earth. A large asteroid, perhaps a dozen miles in diameter, had struck the Earth raising a great cloud of dust and debris which for many months circled the globe blotting out the sun. A thin layer of iridium-rich dust from this event has been found in 65 million year old rocks all over the globe. Iridium, rare on Earth, is much more common in certain meteors and asteroids. The Cretaceous terminal event brought to a close the Cretaceous period of the the Mesozoic ("middle life" era) and heralded in the Paleocene period of the present Cenozoic (or "new life" era.)

During these dark times, the Earth's temperature temporarily lowered and the great dinosaurs died out. Virtually every species on both land and sea of this most successful order perished as well as many other life-forms. Colossalism was no longer the order of the day; the world was abandoned to the small, the furry, and the intelligent.

The life forms which best survived this catastrophe on land were, on the whole, the warm blooded forerunners of modern birds and mammals. While birds like their dinosaur ancestors still laid eggs, most early mammals carried their new born in pouches like modern kangaroos. One of these, a tiny, lemur-like insectovore, probably established the modern line of primates.

Just as the great dying out on the land prepared the way for new terrestrial life forms, so the devastation to sea life led to renewed evolutionary inventiveness. Gone were the great coiled ammonites and reptilian Loch Ness monsters that had formerly ruled the sea. The latter were replaced by a new, brainier, warm blooded mammal which may have reenvaded the sea from the land — the ancient ancestor of modern porpoises and whales. Many mollusks and other simpler forms of sea life which customarily donate their shells to the making of marine fossils were exterminated. Marine fossils are notably absent in the turbidites of Orca Group rocks that were accreting during this period. Plant life fared much better than animal life. Former ferns, conifers, and flowering plants were joined by modern palms, figs and magnolias.

The Cretaceous terminal event had little effect on geological processes at work in the Border Ranges Trench. While new life-forms were taking over the westward moving North American continent, steady accretion continued just off shore of the Superterrane fused to the North American continent. Rocks formed in the accretionary wedge of the Chugach Terrane after the asteroid collision are referred to as the Orca Group in Prince William Sound and as the Ghost Rocks Formation on the western shore of Kodiak Island. Together, these formations are often known as "the Prince William Terrane."

For another 10 million years, earthquakes shook the region, turbidity flows cascaded across the trench floor, and the deposited sediments were carried back into the trench and plastered against its inner wall. As huge quantities of sediments were compressed into accretionary wedges against the landward slope of the trench, it is likely that the trench migrated slowly oceanward and the angle of subduction gradually flattened as it has recently in the modern Aleutian trench off Prince William Sound (Jacobs, 1977). Thus the accretionary prism of the Valdez Group would form the new inner trench wall against which the Orca Group would be accreted helping to explain a number otherwise confusing features in Prince William Sound. First, the Orca deposits seem to be a great distance from the older McHugh trench deposits of the Border Ranges Trench. Secondly, the Valdez Group seems to be composed of mainly slope, inner and middle fan deposits whereas the Orca Group seems to be formed mostly from middle fan turbidites (Winkler, 1976). This is the sequence one would expect for as these sediments are fed into the trench, the earlier subducted deposits would be from the inner portions of the fan and the later from deposits farther out on the fan. However, geologists

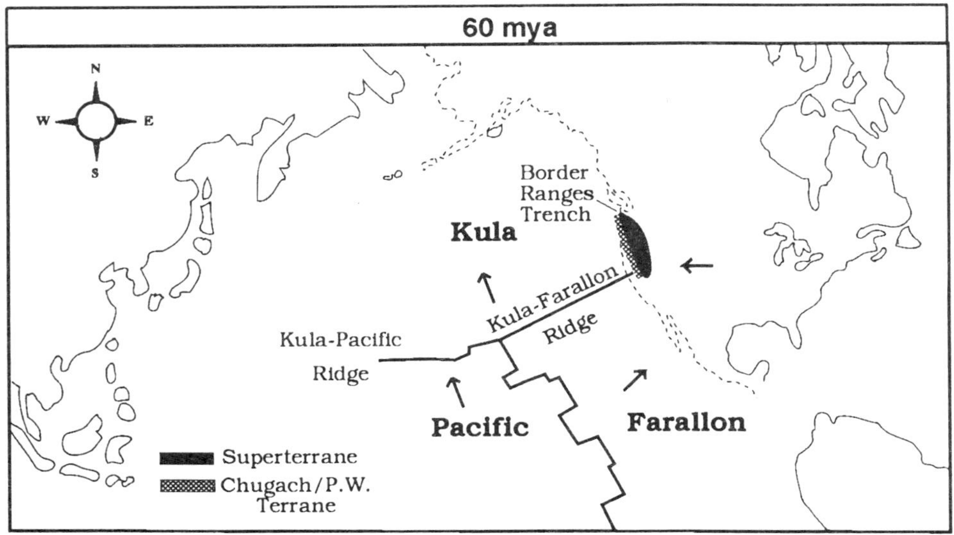

Fig-36. 60 mya. Adapted from Wallace & Engebretson and Gordon & Jurdy

have discovered in the Orca Group, a chaotic jumble of sandstones, conglomerates, and shales which stretches from the vicinity of Miners Lake in Unakwik Inlet to Long Bay near Columbia Glacier. They interpret these deposits as a massive slumping of the inner trench wall (Nelson et. al. 1985). Other, inner fan conglomerates also occur in the Orca Group (eg. the entrance points to Galena and Simpson Bays). The existence of these later, slope and inner fan turbidites alongside the expected middle fan turbidites probably result from the gradual seaward migration of the trench as a gradual build up of accretions extended the inner trench wall seaward.

If the Cretaceous terminal catastrophe was to have little impact on the formation of the Orca Group rocks, another event occurring from about 60 mya was to have a profound effect. The Kula/Farallon spreading center which had been migrating in a northeasterly direction reached the latitude of 40° and began to interact with the Border Ranges trench and the accreting rocks of the Prince William subterrane (Moore et. al. 1983).

As the Kula Ridge approached, the crust beneath the accreting Prince William subterrane began to stretch, thin, and finally pull apart allowing hot basalt to penetrate the wedge and finally pour forth onto the ocean floor. Encountering the cold sea water, the boiling magma would ooze out onto the sea floor, freezing into layer upon layer of basaltic pillows. Turbidity currents cascading down the inner trench wall would occasionally bury these pillowed masses with a layer of sediments which in turn would be buried by extrusions of hot, hissing basalts. These layers of trapped sediments and pillow basalts were then penetrated by feeder vents melting their way up through the pile. These feeder vents on cooling froze into the sheeted dike complex underlying and intruding the pillow basalts of Knight and Glacier Islands. Percolating up through the sedimentary rock the molten basalt would seek out every crack, crevice, and weakness between the bedding planes thus forming the interlayered basalts and sedimentary formations familiar both in Prince William Sound and in the Ghost Rocks formation on the western shores of Kodiak Island.

From place to place, underwater hot springs penetrated the ocean floor spewing forth great clouds of black, sulfide-rich smoke. The sulfides of iron, lead, zinc and especially copper were precipitated out of of this mineral-rich fluid and scattered as thin layers over the ocean floor. Turbidity currents rushing down from the trench slopes scattered thin and sometimes thick layers of sediments over these deposits. At times, these interlayered, mineral-rich precipitates and sediments collected in hollow depressions in the sea floor to be buried by a massive turbidity flow. These formed the lens shaped massive sulfide deposits now found in the Midas Mine, the mines of Latouche Island and at Ellamar.

Other sulfide deposits formed at the interface between oozing pillows; these less extensive deposits can be seen in the many small mine shafts that riddle Knight Island today. In one instance, in the Rua Cove Mine on the eastern shore of Knight Island, geologists may have discovered a sulfide deposit that probably represents the throat of a submarine hot springs.

Knight Island and the Resurrection Peninsula to the west are both recognized as ophiolites formed at a "mid-ocean" spreading center. Both seem to be slivers of mid-ocean ridge, probably the Kula-Farallon spreading center, now emplaced on land. The Resurrection Peninsula ophiolite represents a more complete ophiolitic exposure than does Knight Island, as here the underlying layered gabbro as well as sheeted dikes and pillow basalts have been uplifted. It seems likely that both ophiolites were deposited during the ridge-trench interaction 60 to 50 mya. The Resurrection Peninsula ophiolite has recently been dated at 57±1 million years old (Nelson and Hamilton, 1989) — a date which definitely coincides in time with these postulated events. However, no reliable date has yet been ascertained for Knight Island. Its similar chemical composition and association with Tertiary turbidites probably indicate a similar age. Paleomagnetic results from the Resurrection Peninsula Ophiolite indicate 13° ± 6° (420 to 1140 miles) of movement (Bol, 1987) since its formation.

After their extrusion at the near-trench, offshore spreading center, both ophiolites were probably carried into the trench. As the still hot magma was much less dense than normal, cooled, basaltic ocean crust, the ridge fragments were too light to be subducted. Hence, both were shoved across the trench onto the continental margin and deposited in their present location with respect to surrounding sedimentary rocks. During this process the Resurrection Peninsula Ophiolite was thrust beneath the surrounding sediments.

The Knight Island Ophiolite represents a major structural break in Prince William Sound for the folds in the strata to the east of this feature point toward the northwest (landward) whereas folds to the west point to the southeast (seaward). Seaward vergent folds are normal for most rocks formed in an accretionary prism. However, unusual landward vergent folds are also observed in other areas such as along the modern Washington coast and and off Japan where known active mid-oceanic spreading centers have been subducted at a trench (Byrne and Hibbard, 1987).

Ridge subduction may also help to explain the relative high temperature, low pressure metamorphism of the Chugach Metamorphic Complex just east of the Copper River near Cordova. Typically, sediments in an accreting wedge do not reach a high grade of metamorphism, however, studies of the rocks in this area suggest that they reached 300°C at a depth of about 6 miles within the subducting wedge. Recently, geologists have suggested that Early Eocene subduction of the Kula-Farallon ridge probably provided the heat source for this metamorphic eposode (James et al. 1989).

The anatectic Plutons forming the Sanak-Baranof belt also suggests near trench spreading developed beneath the accreting sedimentary piles of the Chugach/Prince William Terrane. *"Anatectic plutons"* form when sufficient heat is generated within an accreting wedge to melt silica-rich sediments producing a magma having the chemical composition of granite. Basaltic magmas absorbing the silica-rich sediments in an accretionary wedge may likewise form a magma having the composition of granite. The heat source accounting for a belt of anatectic plutons which now stretches from Sanak Island to the west to Baranof Island to the east may have been the same that formed the above ophiolites.

Chemical analysis of the plutons in this belt suggests that they were formed from molten basalts from a "mid-ocean" spreading center melting into the sedimentary wedges of the Prince William subterrane (Moore et al. 1983). There seems to be a gradual younging of these Plutons from west to east — the Sanak Island Pluton is dated at 63 million years, those on Baranof Island about 47 million years. Offsets along transform faults of the Kula/Farallon ridge interacting with the northwestward strike-slipping of this region may account for difference in timing and gradual younging trend of this belt.

Fig-37. Glaciated mountains formed by the Sheep Bay Pluton dominate the skyline above Beartrap Cove.

In Prince William Sound, the Sheep Bay Pluton between Port Gravina and Sheep Bay dates from about 52 mya and is part of this belt. The pluton forming Granite Point between Fairmount and Columbia Bays and lying along Cedar Bay was probably formed at the same time and by the same process as it is chemically similar to the Sheep Bay Pluton. However, no reliable date has yet been determined for this intrusion.

The Sanak-Baranof belt of plutons has important implications for the reconstruction of the geological history of Prince William Sound and Southern Alaska. For example, geologists who believe that the Orca group of rocks were rafted in independently of the Valdez group often use the Sheep Bay Pluton near Cordova to arrive at the date for the latest possible age of emplacement of the Prince William Terrane. Secondly, plutons of this belt overlap the boundaries of the Chugach and Prince William terranes suggesting that at least by Paleocene times, they had to exist as a single unit. Furthermore, one pluton from this belt on Kodiak Island overlaps the boundary between the Peninsular and Chugach terranes establishing that these also were probably joined at least by Paleocene times (62 mya) (Davies 1984).

Paleocene/Eocene Northward Movement of the Superterrane (65 - 57 mya).

Sixty-two mya during the Paleocene geologists place the Kula-Farallon spreading center at a latitude of about 40° (Moore et. al. 1983, Wallace and Engebretson, 1984, Lonsdale, 1988). At this time, it is probable that the Superterrane and its attendant accretionary terrane (the Chugach/Prince William Terrane) were situated near the southern, trailing edge of the Kula Plate. The Superteranne as a part of the continental margin may have been moving northward along with this plate since its establishment 85 mya (Coe et al. 1986, Wallace et al. 1989). Magnetic patterns on the sea-floor of the Pacific ocean reveal that about 56 mya, the common triple junction between the Kula-Farallon, Kula-Pacific, and Pacific-Farallon spreading centers suddenly jumped northward leading to a major reorganization of plate boundaries (Byrne, 1979). As a result, the Kula Plate suddenly began to rotate in a northwesterly direction so that its motion was no longer nearly so perpendicular to the North American continental margin and the Border Ranges Trench. Instead, it began to thrust obliquely beneath the continent. A second consequence of this rearrangement of plate boundaries seems to be that the Kula Plate's velocity with respect to North America increased dramatically (Wallace and Engebretson, 1984).

A major consequence of the sudden counterclockwise rotation and increased velocity of the Kula Plate was to sever the Superterrane from the North American Continental Margin along its former suture zone (the Gravina-Nutzotin Terrane) so that it began moving rapidly northward with the Kula plate (suggested in part by Lonsdale 1988). The Border Ranges Trench continued to accrete the sediments of the Orca Group; however, the new, oblique convergence caused volcanism in the Superterrane to cease abruptly 56 mya (Wallace and Engebretson, 1984). The Border Ranges Trench then seems to have moved northward with the collage of terranes aboard the Kula Plate.

Thus, after 53 million years the Superterrane was once again moving north — now astride the rapidly moving Kula Plate, strike-slipping abrasively against the North American margin. But the Superterrane had assumed a wholly new aspect from the composite terrane which had approached the North American continent so many tens of millions of years ago. On approaching the continent, it had accreted the Gravina-Nutzotin flysch/magmatic belt and the Chugach/Prince William accretionary terrane had formed along the Border Ranges Trench. On tearing away from the continent, a sizable piece of Wrangellia was left behind attached to the North America in the Hells Canyon region of Oregon. And where Wrangellia and the Alexander Terrane had been welded to the continent by the Coast Plutonic complex a large piece of the continental margin, itself a previously accreted terrane now known as the "Stikine Terrane," was probably sliced off to accompany the Superterrane on its strike-slipping journey northward (Coe et. al. 1986).

Although we cannot reconstruct the exact configuration of the Superterrane 60 mya from the various pieces now strewn out along the coast from British Columbia to Southcentral and Southwestern Alaska, we have at least a rudimentary grasp of the forces that created such a disarray. When a terrane strike-slips along a continental margin at an obliquely convergent angle, several things may happen. It may be elongated by being slivered along parallel transform faults; parts of the terrane may adhere to the continental margin to be left behind by the more rapidly moving outboard sections; the terrane may be fragmented during the collision and these fragments rotated.

The Eocene Flight Northward (56 - 43 mya)

During the Eocene (57-37 mya), as the submerged Chugach/Prince William Terrane, attached to its source terrane, moved northward along the continental margin, the Earth's climate by modern standards remained generally temperate but began to exhibit a long term cooling trend. Lush rain forests consisting of giant redwoodlike conifers spread over the moister areas of North America. It was at this

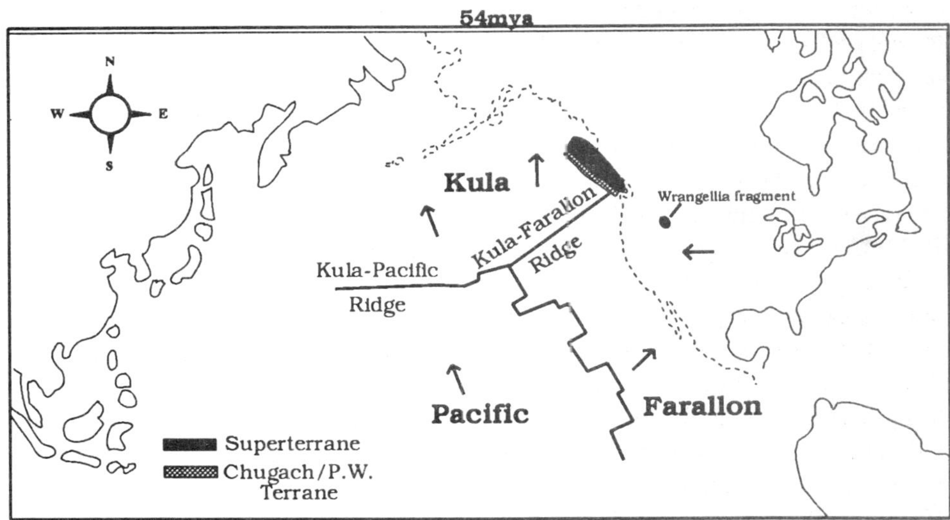

Fig-38. 56-54 mya.

time that the mammals were to establish themselves as a dominant life form. Early rodents, bats, and whales emerged. The *eohippus* or "dawn horse," a tiny creature the size of a modern dog, sporting four horny toes on its forelegs and three on its rear legs, served as a prototype of the new hoofed mammals. In Africa small elephant-like mammals, the size of modern pigs, emerged. The great terror of the Eocene mammals was a giant, carnivorous ostrich-like bird that recalled its dinosaur past, the swift and deadly *Diatryma*. Taking refuge in the trees, a small primate with grasping hands and forward looking eyes was able to elude this terrible predator preparing the way for the orders of monkeys, apes, and finally man.

Because the Superterrane was riding on a runaway Kula Plate traveling northward at almost 10 inches/year, tremendous amounts of energy were released as its eastern edge underthrust and scraped past the continental margin. Great metamorphic and plutonic belts now mark this collision zone in present day British Columbia and Southeast Alaska and major magmatic events occurred all along this coast between 54 and 45 mya (Brew and Morrell, 1983).

For roughly 11 million years, the Superterrane and attendant Chugach/Prince WilliamTerrane lurched violently northward. During this process, the terranes were fractured along innumerable longitudinal strike-slip faults. Slivers shifted haphazardly and haltingly past one another. Dismembered fragments would rotate crazily against one another and the continental margin. Occasionally, pieces of the continental margin would be torn loose to travel northward with the advancing terrane. Fragments large and small would sometimes cling to the continent to be left behind by the advancing amalgam of splintering terranes. Vancouver and the Queen Charlotte Islands represent one large fragment of Wrangellia left behind. Another small detached splinter of Wrangellia can be found stretching northward from Sitka to the Yukon Border. A small sliver of the Chugach Terrane can be seen today, still clinging to the Vancouver Island fragment of Wrangellia (Cowan, 1982). A piece of the Alexander Terrane came to rest scissored between fragments of Wrangellia and the Tracy Arm Terrane. The Gravina-Nutzotin Terrane has been drawn out into two, thin oblong fragments — one resting in the Yukon Territory and the other in southeastern Alaska. The outboard Peninsular and the Chugach/Prince William Terranes were stretched out in thin arcs but, in general, seem not to have suffered the fragmentation of the colliding, inboard terranes. Thus the violent but slow forces of oblique collision created the crazy quiltwork of terranes known as Southeastern and Southcentral Alaska.

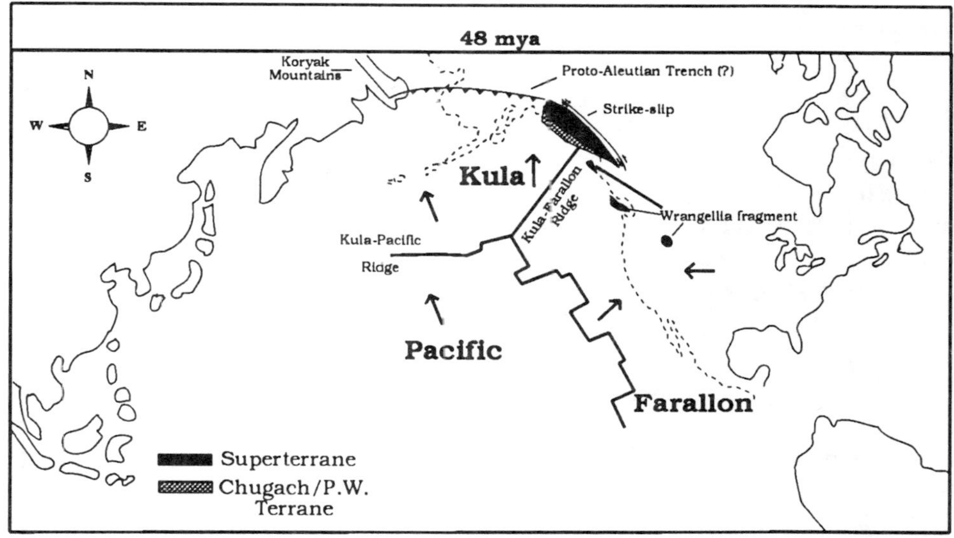

Fig-39. 48 mya.

The Final Collision (48 - 43 mya)

While the inner, eastern edge of the Superterrane was interacting with the continental margin, the northwestern edge was closing a large ocean basin that had bordered the old Mesozoic margin of proto-Alaska at least since Jurassic times. Here, the ancient forerunner of the modern Aleutian trench slowly accreted a flysch terrane as it steadily consumed the northward advancing Kula Plate. The trench at this time stretched from the Koryak mountains of Siberia and along the Mesozoic margin of proto-Alaska. The forward edge of the Superterrane, which probably consisted of interlocking sections of Wrangellia and the Peninsular terranes, acted like a giant bulldozer crumpling the Jurassic/Cretaceous flysch of the ocean basin before it. In advance of the Superterrane a number of small, exotic crustal fragments (miniterranes such as the Chultna Terrane) were traveling toward the Mesozoic trench. These were squeezed in the compressed flysch, which was crumpled between the continental margin and the rapidly advancing superterrane, and were bulldozed across the proto-Aleutian Trench onto the continental margin (Csejtey et. al. 1982). J.C. Moore (1983) suggests that flysch formed in Paleocene-

Eocene times may have been buried by the docking Superterrane which overthrust the crumpled flysch zone for a distance of 60 to 120 miles (Csejtey et. al,1982). The great forces created by the late Paleocene/early Eocene approach of the Superterrane squeezed the flysch in the closing basin into great wave-like folds (cf. Fig-20). So great was the energy released in this folding process that melting occurred in the silica-rich sediments fusing great masses of them into granitic plutons. One of these is responsible for the present edifice of Mount McKinley (Coney et al. 1985). In fact, like the Himalayas which are the result of a collision of India with the continental mass of Asia, the present day spectacular uplift of the central Alaska Range may be due in part to a micro-continental collision of the Superterrane with proto-Alaska.

It is difficult to say exactly when the Superterrane and the attendant Chugach/ Prince William Terrane reached their present locations. Recent paleomagnetic studies in both the Peninsular and Wrangellia terranes suggest that portions of the Superterrane were in place by at least 50 mya (Hillhouse et al. 1985, Thrupp and Coe, 1986).[6] Undoubtedly, the final docking of the strike-slipping Superterrane and the shortening of the trapped flysch basin occurred over many millions of years, but the final closing of the proto-Aleutian trench probably occurred during an 8 million year period from 50 to 42 mya. This event seems to be indicated by a number of lines of evidence. First, the Chugach/Prince William Terrane attached to the outboard edge of the Superterrane seems to have been in its present position by a least 42 mya (Keller et al. 1984, von Huene et al. 1985). It is conceivable, considering its most outboard position, that this terrane may have ended its strike-slip journey north sometime after the Wrangellia portion (Coe et al. 1985, cf. [6]).

Several events dating 43-42 mya suggest that the final docking of the Superterrane and the consequent jump of the Aleutian Trench to its present position may have had profound consequences for the entire North Pacific Basin (Lonsdale, 1988). First, rapid spreading along the Kula/Pacific ridge suddenly ceased. In fact, spreading along the ridge seems to have died out completely; so that there remained one giant ocean plate dissected in an east-west direction by a cooling and no longer active mid-ocean ridge creating a new, composite Kula-Pacific Plate. In 1986, scientists studying the ocean floor off Attu in the Aleutian Chain discovered the final fragment of the fossilized Kula-Pacific spreading center still subducting in the Aleutian Trench (Lonsdale, 1988).

Plate reconstructions tell us that 43 mya, the new composite plate shifted its direction of movement farther to the northwest — the direction which characterizes the present Pacific Plate. This change of direction of plate motion is recorded as a sharp bend in the hot spot track of the Emperor seamount chain. The demise of the Kula-Pacific spreading center also resulted in a significant slowdown in the

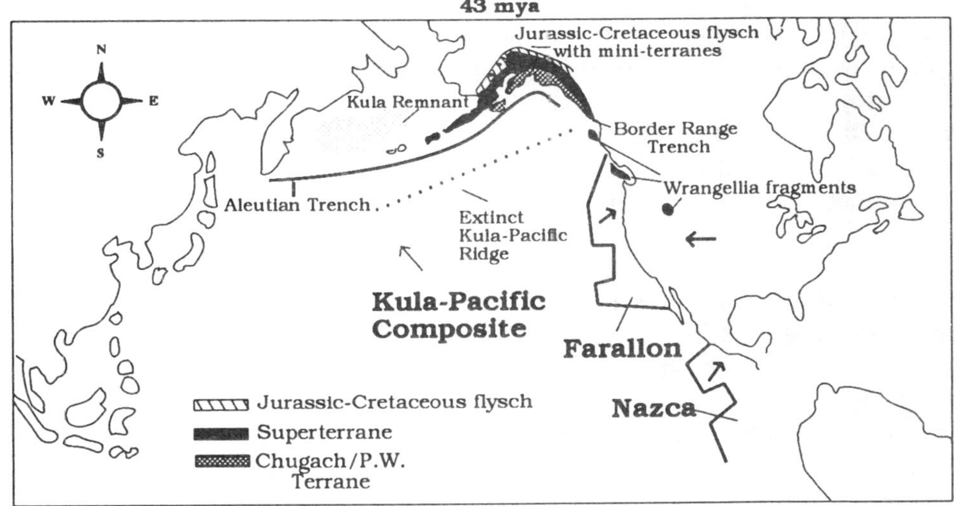

Fig-40. 43 mya.

rate of plate movement and hence in the rate that the plate was being consumed beneath the Alaskan margin; this rate slowed from a breathtaking 10 inches a year for the Kula Plate to the present rate of a little over 2 inches/year for the Kula/ Pacific composite plate (Wallace and Engebretson, 1984).

Not only does the date of 43 mya mark a possible date for the emplacement of the final fragments of the Superterrane and the end of a period of significant plate reorganization which began 56 mya, it also probably dates the establishment of the present Aleutian Trench (Wallace and Engebretson, 1984). Paleomagnetic studies indicate that the Aleutian Arc, unlike the attached Alaska Peninsula and its out-board accretionary terranes, has undergone little or no significant northward movement (Harbert, 1987). Geologists believe that an older proto-Aleutian trench stretched from the Koryak Mountains of Siberia across the Bering sea region and then along the Mesozoic margin of Alaska. Sometime in the Paleocene or Eocene, it is thought to have jumped to near its present position (Cooper et. al. 1976, Lonsdale, 1988). Evidence for a jump in the Aleutian Trench lies under the Bering sea in the Aleutian Basin just to the North of the present Aleutian Island Arc. Here, the sea floor of this marginal basin is uncharacteristically older than the adjacent arc suggesting that the basin is a remnant of the Kula Plate trapped behind the arc when the proto-Aleutian Trench jumped from the Koryak Mountains to its present position (Cooper et. al. 1976).

Because the present Aleutian Trench marks a convergent zone between ocean crust and ocean crust on its western end but between ocean crust and continental crust on its eastern end, it is tempting to speculate that, like the Bering Sea Basin, the Peninsular Terrane, attached Chugach/Prince William Terrane, and portions of Wrangellia were also isolated as a part of the continental plate by this same trench jump and that the collision of the microplate itself was the probable cause of the relocation of the trench (cf. discussion in Ben-Avraham, Z., and A. Nur. 1983).

Forty-three mya also marks the beginning of renewal of magmatism in Southern Alaska ending the 13 million year lull that occurred as the Superterrane began strike-slipping northward. However, it should be noted that magmatism in Southern Alaska before 56 million years ago was probably associated with the Border Ranges Trench at lower latitudes; whereas magmatism after 43 million years ago is probably associated with the present, Eocene Aleutian Trench.[7]

When the northern, leading edge of the Superterrane collided with Alaska, the continental margin most likely had a broad arcing form caused by compression between the eastward moving Eurasian plate and the westward moving North American Plate that occurred 67-53 mya (Coe et. al. 1986). However, the the sharper oroclinal bending of the more recently arrived terranes in the Prince William Sound region may be due to the interaction of strike-slip and convergent forces at work at this juncture. Paleomagnetic studies indicate that terranes in this region have undergone considerable counterclockwise rotation — perhaps caused by strike-slip forces acting on the inboard edge of stalled, obducted blocks. These forces acting for many tens of million years (probably into present times) have tended to create broad arcing faults inland and to smear the already elongated crustal slivers around the "oroclinal" bend giving the terranes of Southcentral Alaska their peculiar arcing shapes.

Oligocene Plutonism (37-24 mya)

Although most of Alaska had accumulated near its present position by 40 mya, a late Eocene/early Oligocene Alaskan map would have differed from modern versions in a number of distinctive ways. Only a few of the youthful volcanic Islands of the Aleutian arc might be seen poking through the surface of the Pacific Ocean. Kodiak Island, the Kenai Peninsula and the Prince William Sound area had not yet been uplifted. Although these had been transferred to the continental plate by the jumping of the proto-Aleutian trench, it is likely that they existed as a part of the submerged continental shelf. The stretch of coastline between Prince William Sound and the Inside Passage presently occupied by the town of Yakutat would also be noticeably absent as it had not yet arrived aboard the Pacific Plate.

Thirty seven mya, life on earth suffered another in a series of ecological catastrophes. Many marine and terrestrial species suddenly disappeared from the face of the earth. Once again a cosmic cause may be indicated as millions of tiny, glassy droplets of molten rock called "tektites" were dispersed over half the planet, probably as a result of the impact of a great meteor. This event marks the end of the Eocene and the beginning of the Oligocene (37-24 mya).

The Eocene terminal event seems to have altered the Earth's climate marking a long-range plunge in global temperatures that continues to the present. Whereas, previously, July and January temperatures had not differed markedly; now suddenly, winter had once again become a fact of life. The already successful mammals with their invention of fur and warm bloodedness found the cooling temperatures very much to their liking as did the warm blooded and active birds which added many new species to their expanding numbers.

By early Oligocene times, the young Aleutian Trench had established itself off the continental shelf and began to generate magma. Plutons intruded the Alaska Peninsula and volcanoes erupted in the Kula plate remnant to the Southwest building the Aleutian Island Arc. Near Prince William Sound, the newly formed trench probably dipped down at a much steeper angle than today such that portions of the Chugach Terrane lay above the 50 mile-deep zone of partial melting. As the magma forced its way toward the surface, it slowly cooled, crystallizing into granite plutons. These 35 million year old plutons subsequently uplifted and unroofed by glaciers form the bulk of Culross, Esther, and Perry islands, the granite pluton bordering Passage Canal and Port Wells, the granite domes of the Nellie Juan/Kings Bay area and the Eshamy Bay granitic outcrop. Metamorphosed rock lying between the Culross Island, Nellie Juan and Eshamy plutons suggest that they may represent the surface exposure of a single large pluton (Tysdal and Case, 1979). So intense was the heat during intrusion that contact metamorphism baked the sedimentary rock surrounding these bodies to a distance of from one half to two miles.

The pluton underlying the Passage Canal and Port Wells area created considerable mineralization. Here, one can observe mineralized quartz veins which were injected into the surrounding sedimentary rock as the pluton slowly cooled.

Just as renewed magmatism in the recently docked terranes resulted from subduction at the newly formed Aleutian Trench, so a sedimentary wedge began to form along the new Alaskan continental margin. Thus, the late Eocene-Oligocene Sitkalidak Formation of the Kodiak Islands is easily distinguished from the older Paleocene Ghost Rocks Formation formed during the ridge-trench interaction at 40° latitude (Moore et al. 1983). Its apparent counterpart in Prince William Sound, the Eocene Montague belt, is likewise easily distinguished from the Bainbridge melange farther inland (Helwig and Emmet, 1984).[8] Sedimentary wedges of Oligocene age in the Prince William Sound area either have not yet been uplifted or have been removed by erosion. The steady accretion which began in the late Eocene continues in the Gulf of Alaska today.

Miocene Uplift (24 - 5.3 mya).

The Miocene (24-5.3 mya) was marked by a fluctuating but gradual deterioration of the climate resulting in the alternate raising and lowering of sea levels as the polar ice caps waxed and waned. Land bridges exposed during periods of low sea level encouraged the continental exchange of mammals and other orders. The great evolutionary invention of the Miocene was grass. Evolving from flowering plants, grasses proved a hardy species during cooler, drier periods when much of World's moisture was locked up in the polar ice caps. Moreover, in response to volcanically caused fires, natural selection in grasses developed a new survival strategy : rather than growing from the leaves down like most plants, grasses began growing from the roots up. Not only did this adaptation render them impervious to fire, but provided a perpetual harvest to the grazing mammals and the carnivores which fed upon them. Needless to say there was a virtual Miocene explosion of herbivores.

Meanwhile, our ancestors were making steady but slow progress toward the present. By 35 mya the forerunners of the New World monkeys had developed in South America while 29 mya their Old World counterparts roamed the lush forests of Egypt. By 20 mya a half-erect tailless ape inhabited Africa. But it was not until the end of this period, about 5 mya, that our early ancestors took on the most primitive of human traits.

At the beginning of the Miocene (24 mya), the Alaska Peninsula and the Kodiak-Prince William Sound area still existed as submerged portions of the Southern Alaska continental margin. Sometime around 20 mya, the leading edge of the northward moving Yakutat terrane collided with the Aleutian Trench near the present Prince William Sound region (Plafker, 1988). This subduction process was to initiate a period of uplift which would regionally metamorphose, deform and elevate the entire region, creating a temporary lull in magmatic activity in the Alaska Peninsula but activating the volcanoes of the Wrangell Mountains. The deformation (faulting and folding) associated with this event climaxed in the Orca Group rocks about 15 mya and probably continues into the present (Helwig and Emmet, 1984).

Metamorphism in the Valdez Group during this period of uplift produced much of its present mineral wealth. The gold and silver of the prosperous Cliff Mine and other mines of the Valdez-Columbia Glacier area were probably formed at this time. So intense were the pressures of uplift that they contorted the rock into accordian-like folds releasing great quantities of heat. Silica-rich fluids were sweated out of the sedimentary rocks carrying with them silver and gold. Thus, these valuable minerals which had been distributed evenly throughout the Valdez turbidites became concentrated. It may be that the turbidites in the Valdez region were particularly rich in gold due to nearby undersea volcanism in the Eocene. The

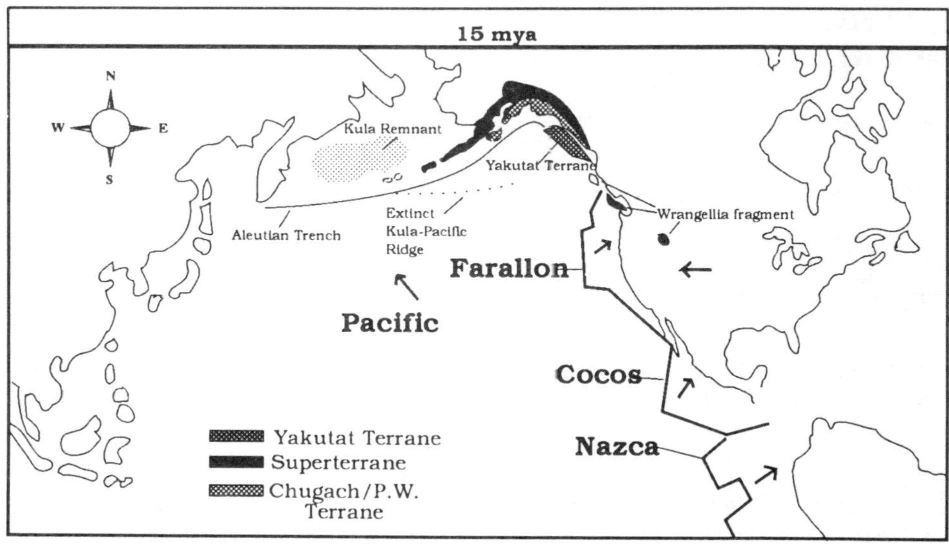

Fig-41. 15 mya

prosperous Cliff Mine (production 51,740 oz. gold) lies only 14 miles from the Ellamar Copper mine whose sediment hosted sulfide deposits produced 59,874 ounces of gold. On the other hand, only minor amounts of gold (484 oz.) were produced from the sulfide deposits on Latouche Island (all production figures here and ff. are from Jansons et al. 1984). There are only a few minor gold deposits in the southwest corner of the Sound.

The same forces of uplift which folded some of the rock shattered the less pliable strata. The mineral-laden quartz fluids were injected into these fractures crosscutting the preexisting bedding planes. Most of the gold in this region is found in quartz veins which crosscut the strata while quartz veins parallel to the bedding planes, probably associated with the original accretion of these rocks, are mostly barren (Fechner and Krause, 1987).

Continued Pliocene Uplift (5.3 - 2 mya).

Five mya at the beginning of the Pliocene, when our early ancestors in Africa had barely become recognizable as human prototypes, continued underthrusting of the leading edge of the Yakutat terrane initiated the present uplift of the Kenai Mountains, which before this time may have existed as an eroded lowland. Because high mountains in the gap between a trench and its volcanic arc are rare, scientists suspect that the Kenai mountains, including the island peaks of Prince William Sound and the mountains of Kodiak Island, were and still are being uplifted by the collision of the Yakutat Terrane (Von Huene et al. 1986). The general uplift in this area, which continues today, is not simple; for the terrific compression caused by the accreting terrane manifests itself in a wave-like manner — downwarping certain areas while uplifting others. For example, the drowned fiords and cirques of the Kenai Peninsula and Prince William Sound may be a sign that these areas are presently temporarily subsiding. Along the eastern edge of the terrane in the Yakutat area, the uplifted coastline and high mountains of the Fairweather and St. Elias Ranges seem to be evidence of oblique underthrusting of the continental margin by the accreting terrane.

Although the Kenai and Chugach ranges were uplifted by 5 mya, we should not imagine the Prince William Sound area as resembling the ragged mountains and deep fiord system we know today; for the great glaciers had not yet arrived to sculpt these mountains into their present configurations. Most probably the Chugach Range existed as an uplifted coastal plateau cut by v-shaped valleys formed by numerous streams descending to the sea. Some of these valleys would later provide the pathways for glaciers which would scour out the deep fiords of the Sound.

Pleistocene Glaciation (2 - .012 mya).

Sometime, about 2 mya, average summer temperatures became slightly colder than they are today; winter snows no longer completely melted so that gradually, year by year, they compacted into solid ice. Great glaciers began to fill the valleys of high and temperate latitudes until finally, they joined into huge ice caps which soon covered most of North America, much of Northern Europe and Siberia. While a great ice sheet spread over most of Southern Alaska, Prince William Sound and the Chugach/Kenai Mountains, much of interior Alaska remained ice free — probably because, like today, great walls of mountains intercepted the moisture-laden air flowing from the Gulf of Alaska leaving an arid but frozen desert in their lee.

The ice ages of the Pleistocene were by no means constant; instead the ice caps waxed and waned and glaciers advanced and retreated. According to how far south the glaciers advanced, American scientists distinguish four major glacial episodes for North America — the Nebraska, Kansas, Illinois, and Wisconsin periods. Between these episodes were three milder interglacial periods in which temperatures and the extent of glaciation may have been similar to those of the present interglacial period.

Undoubtedly, Southern Alaska and Prince William Sound participated in these major fluctuations so that ice alternately embraced and exposed the area. Unfortunately, each glacial advance has obliterated the evidence of each former advance; thus, little is known about the nature and extent the Nebraskan, Kansan, and Illinoisan glaciations in Southern Alaska. Probably, most of the area existed under giant ice sheets during periods of world-wide advance and emerged from under the the ice during interglacial periods. Each new advance of the ice must have scoured out new valleys, rounded off the peaks of former mountains, bulldozed away former moraines and possibly whole ridge systems. When the area lay under a great burden of ice, the crust below was pressed down into the mobile asthenosphere offsetting tectonic uplift of the region. During these periods, because so much water was locked up in glacier ice, the sea level would fall exposing presently submerged portions of the continental shelf such as the famous Bering land-bridge. Near the end of the Pleistocene, this landbridge provided the migratory route for many animal species that now populate North and South America. Indeed, early American Indians arrived via this same route.

During the interglacial periods, the glaciers would shrink leaving behind great arc-shaped piles of rubble known as "moraines." As the ice melted, sea levels would rise once again covering the former continental shelf and inundating low lying valleys. Such a sequence of events was to repeat itself several times during the early and middle Pleistocene.

It was not until the last glacial advance of the Wisconsin age climaxing about 20,000 years ago that Prince William Sound began to assume its present form. At this time, great tongues of ice flowed down from the uplifted Kenai/Chugach Mountains to pool over the Prince William Sound region into a massive ice sheet known as a "piedmont glacier," perhaps similar to the present Malaspina Glacier. Just west of Naked, Peak, and Storey Islands, this mile-thick accumulation of ice scoured out one of the deepest depressions in the present American continental shelf. Here a submarine trough lies 2,400 feet below today's sea level. The weight of this great ice mass may have depressed the crust 480-864 feet (Lethcoe, 1987). A number of major glaciers fed this massive ice sheet which probably extended out onto the present continental shelf at least as far as Middleton Island. A 3,200 foot thick river of ice flowed out of what are now Port Valdez and Valdez Arm (Grant and Higgins, 1913). This glacier in turn was fed by minor tributaries from Valdez Glacier and the glaciers occupying Jacks and Galena Bays. Shoup Glacier, now barely tidal, is all that remains in the Sound of this once great tributary system. To the south and east, glaciers from Port Fidalgo and Port Gravina further contributed to this ice stream. To the west where the high Chugach still accumulates enough snow to nourish the glaciers, one can observe today the remains of the last tidewater tributaries that joined the Valdez system during the Wisconsin Ice Age. Although much diminished in size and extent, Columbia and Meares Glaciers still push down from the Chugach mountains as they must have for the past 20,000 years.

Flowing south, this great tributary system overrode the former mountain ridges which now form the Naked Island group, Smith, Seal, and Green Islands. Farther south, the ice encountered an elongated ridge system perpendicular to its direction of advance — the rocks forming present-day Hinchinbrook and Montague Islands. The advancing ice for the most part overrode these ridges rounding off the once lofty summits. Where Hinchinbrook entrance lies today, the gigantic ice mass furrowed out a deep U-shaped trough to join the large ice sheet that spread out over what is the present day continental shelf.

A second, massive tributary system flowed down from the northwest. Port Wells, Harriman and College Fiords were carved out by a great valley glacier perhaps 4,000 to 5,000 foot thick (Tarr and Martin, 1914). The present tidewater glaciers of Harriman and College Fiords, impressive as they are today, represent only the final, fading remnants of this once great system. To the west, the Port Wells system was joined by rivers of ice issuing from Passage Canal, Blackstone Bay and Port Nellie Juan. The combined ice stream from all these glaciers gouged out a deep trough now known as Knight Island Passage. This southward moving stream, abutting the Knight Island ophiolite and the uplifted bulk of the Montague out a deep trough now known as Knight Island Passage. This southward moving stream, abutting the Knight Island ophiolite and the uplifted bulk of the Montague ridge, was diverted to the southwest carving out Montague Strait, Latouche, Elrington, Prince of Wales and Bainbridge Passages as it exited the southwestern corner of the Sound.

It is likely that this second major tributary joined the first, flowing from the northeast around the southern end of the Knight Island ophiolite which stood like like a giant, resistant boulder in an eddying river. Although peaks on the northern end of the Knight Island group were overridden and rounded off by the ice, several jagged peaks over 3000 feet high on the southern end must have thrust above the ice sheet as lonely nunataks.

Holocene Retreat (.012 - Present).

About 12,000 years ago, for reasons not completely understood, the earth's climate began to warm dramatically. Evidence gathered from the measurement of ancient peat bogs in Prince William Sound suggest that much of the area had lost most of its ice-cover by about 10,000 years ago. If so, the retreat of the ice must have been drastic indeed. The present retreat of Columbia Glacier, although minuscule by comparison, might provide a modern analog to what may have happened so

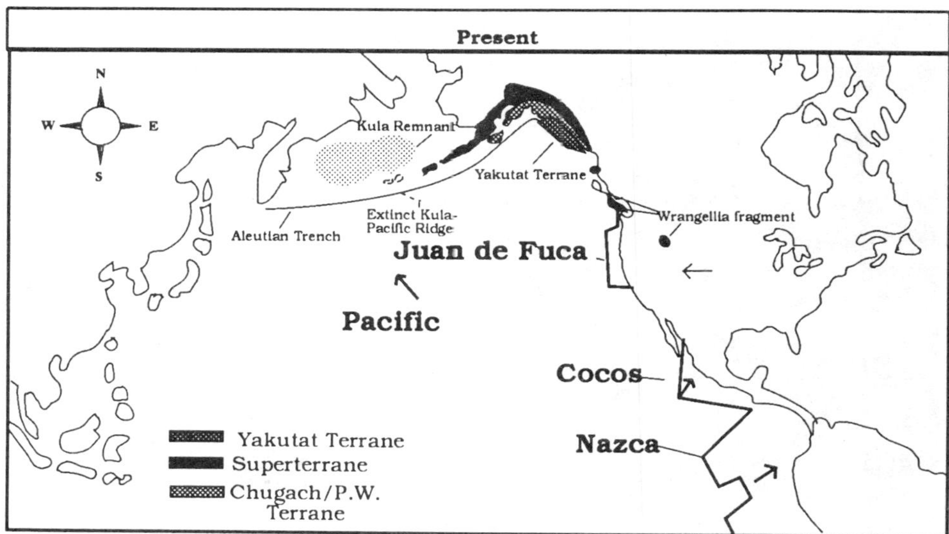

Fig-42. Present.

many thousands of years ago. Like Columbia Glacier which has retreated off its moraine into a deeper basin exposing more of its terminus to the melting effects of warm seawater, the Pleistocene ice sheet poised on the shallow continental shelf of the Gulf of Alaska also retreated back into the deep basin it had carved in the Prince William Sound area. Hundreds of feet of ice face were exposed to the warming effects of the invading and rising seas leading to a rapid breakup of the ice covering the basin.

The drastic retreat of the ice sheet and the continuing fluctuation of the retreating glaciers during the last 10,000 years (the Holocene) had several important large scale effects on the topography and bathymetry of the Southern Alaska coastline. The Earth's crust relieved of the heavy overburden of ice began to rebound elastically. Although it is difficult for geologists to distinguish this isostatic rebound from tectonic uplift created by subduction in the Aleutian trench, at least several hundred feet of uplift of the coastal ranges are probably attributable to this cause. The Influence of this uplift on the shape of the Southern Alaska coastline has, however, been balanced to some extent by a global rise in sea level of some 300 feet resulting from the melting ice.

Like former glacial advances and retreats in this area, which may date as far back as the Miocene (Molina, 1986), the present retreat has resulted in the introduction of great loads of sediment onto the continental shelf and thus into the Aleutian Trench through the action of turbidity currents. The present thick

Fig-43. The retreating ice left behind a rugged coastline of fiords and glacially over-ridden islands. Blackstone Bay. Photo courtesy of Nancy Simmerman.

accretionary wedge jamming the trench may be partly responsible for the shallow angle of subduction in the Prince William Sound area and the consequent violence of the 1964 Earthquake.

The retreating ice left behind the glacially carved landscape of Prince William Sound that we know today — a rugged mountain landscape of Matterhorn peaks, saw-toothed ridges, steep-sided, hanging valleys and amphitheater-like cirques. What the ice had begun, the invading Pacific completed. Where former rivers of ice had gouged U-shaped troughs in the bedrock, the invading sea formed an intricate system of deep fiords; where tributary glaciers once fed the larger streams, drowned valleys formed the Sound's many bays and passages; where the ice relentlessly ravaged the mountain slopes, a ragged shoreline of smaller sheltered coves and harbors now stands.[9]

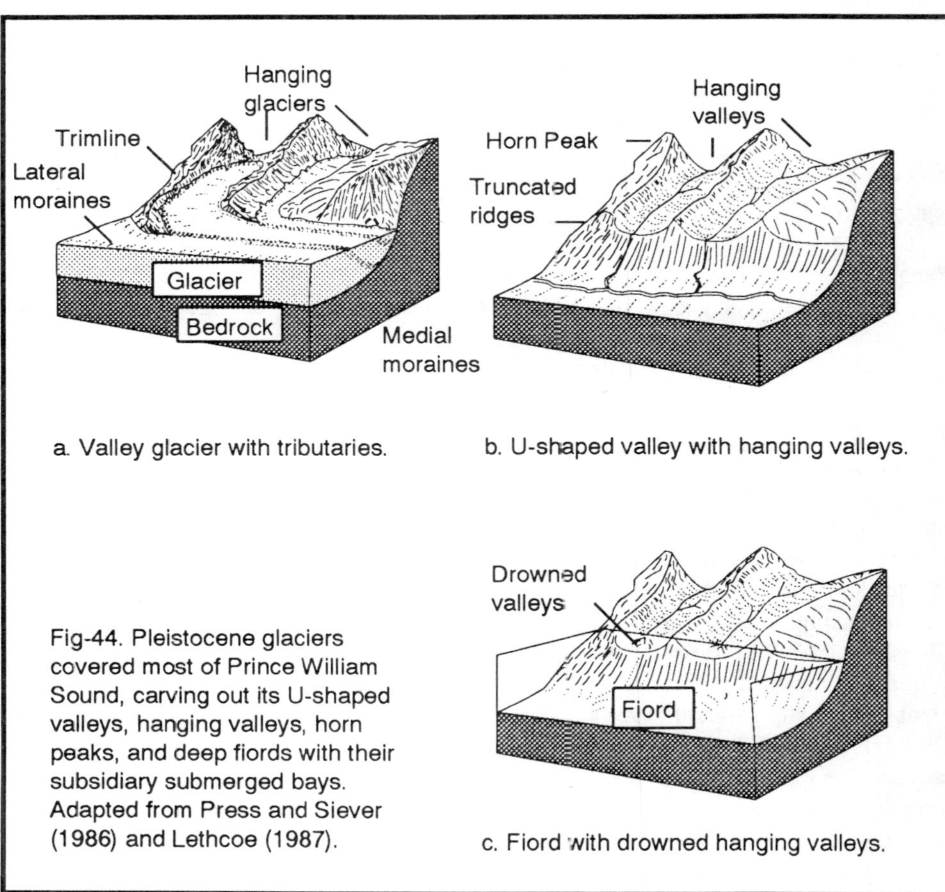

a. Valley glacier with tributaries.

b. U-shaped valley with hanging valleys.

c. Fiord with drowned hanging valleys.

Fig-44. Pleistocene glaciers covered most of Prince William Sound, carving out its U-shaped valleys, hanging valleys, horn peaks, and deep fiords with their subsidiary submerged bays. Adapted from Press and Siever (1986) and Lethcoe (1987).

Footnotes

1. The proposed scenario in the Geological History section to follow is based on the following assumptions and conclusions from other studies:

a. Because rocks on either side of the Contact Fault are difficult to distinguish and because no suture zone is distinguishable in Western Prince William Sound, the rocks of Prince William Sound are assumed to compose a single terrane.

b. Rocks of the Prince William Sound area seem to be the result of continuing accretion lasting from the late Cretaceous through the mid to late-Eocene. We will, however, use the traditional terms "Valdez" and "Orca" to refer to the older and younger rocks of this single accretionary prism. (Cf. Introduction above).

c. Rocks of the Chugach/Prince William Terrane seem to have been formed at a Paleo-latitiude of about 40° North which is approximately the same latitude as the southern-most fragment of the Superterrane (Hell's Canyon).

d. A Jurassic microfossil found in the McHugh formation near Anchorage suggests affinities with fossils found in Oregon (Nelson et. al. 1986) and Cretaceous ammonite fossils recently discovered at Ziegler Cove in Prince William Sound appear to correspond to fossils found in the Pacific Northwest (this study).

e. The Superterrane is thought to have been formed at a more southerly latitude, collided with the North American continent and possibly moved northward along the continental margin through strike-slip faulting (Jones et. al. 1977, Coe et al. 1985).

f. Rocks of the Chugach/Prince William Terrane seem to have been formed in the Border Ranges trench which itself appears to have been displaced northward (Jones et. al. 1978).

g. Rocks of the Chugach/Prince William Terrane are presently contiguous at the Border Ranges Trench to all three subterranes of the Superterrane — although it is not certain how these relationships have been altered over time through strike-slip faulting and telescoping. For example, cobbles containing Devonian corals have been found in the Chugach Terrane presently attached to Wrangellia near Valdez (this study). The only source terrane old enough to have supplied these is the Alexander Terrane whose closest attachment to the Chugach Terrane is some 260 miles east of this location. Also, rocks apparently from the Strelna formation of Wrangellia appear in the McHugh Complex near Anchorage presently attached to the Peninsular Terrane (Nelson et. al. 1986). The Raspberry schists (blueschists) of the Peninsular Terrane on Kodiak Island are thought to have possibly been been translated along the Border Ranges Fault system from Wrangellia on Vancouver Island (Roeske et. al. 1989). However, George Plafker believes that "Sutures between the Prince William, Chugach and Wrangellia-Peninsular terranes have had no large post-emplace-ment dextral slip (Plafker, 1988)." Large displacements might not be expected if the Chugach/Prince William accretionary terrane traveled northward with its source terrane, however the studies cited above and Cowan (1982) cited below suggest that inter and intra-terrane displacement may have indeed been significant. Coe et. al. (1985), for example, after reviewing the literature on fault displacements in Southern Alaska feel that 20° of displacement is not unreasonable.

h. A fragment of the Chugach Terrane, whose rocks match those south of Sitka, has been discovered adhering to a fragment of Wrangellia at the latitude of southern Vancouver Island (Cowan, 1982) indicating that these two terranes were linked while at a latitude, at least, this far south.

i. Maximum uplift of the coast metamorphic and plutonic belt occurred between 47-52 mya (Hollister, 1979) — probably due to underthrusting of the Superterrane on its strike-slip journey northward.

j. The Peninsular Terrane, because it represents a magmatic arc eroded to its plutonic roots, provides a credible sediment source for the formation of the sandstones of Prince William Sound. (Dumoulin, 1987).

k. Rocks of the Kodiak Ghost Rocks formation of the Prince William subterrane seem to be derived from both continental and island arc sources (Moore et, al. 1983). The same seems to be true for the Orca group (Plafker,1988).

l. If the Chugach/Prince William Terrane has been displaced northward by 20°, the easiest explanation for the uniformity of sandstone compositions in the Sound is to assume that it traveled northward attached to its source terrane — most probably the Peninsular Terrane (Dumoulin, 1987).

m. Sixty-two million-year-old plutons on Kodiak Island have been discovered stitching together the Prince William, Chugach and Peninsular terranes (Davies and Moore, 1984). Since this age is very close to the age of formation of the Prince William rocks, there seems little reason to assume that the Prince William and Chugach terranes ever existed as totally distinct entities with different travel histories. By at least 62 million years ago, they were probably attached to and moving with the Peninsular Terrane. Geological evidence linking the Superterrane to the Chugach/Prince William Terrane before Eocene times is offered also by Little (1988) and Pavlis (1982). Finally, a study of Orca group (Prince William Terrane) sandstones along the TACT route and on Montague Island (Gergen and Plafker, 1988) concludes that the Orca sandstones may have been derived from erosion of older Valdez (Chugach Terrane) rocks suggesting a linkage of these two groups over time.

2. For an alternate scenario see (Plafker, 1988. and Plafker et. al. 1989)

3. No one knows exactly the configuration of ocean plates in the proto-Pacific during Jurassic times, for we are dealing here with a period as old as the oldest ocean rocks themselves. All evidence from these rocks has since been destroyed in the subduction zones fringing the ancient ocean. Hence, this scenario, while plausible, is purely speculative but does account for a number of events geologists think probably occurred in the mid-Jurassic, around one hundred and fifty million years ago. First Wrangellia which probably had been drifting on its ocean plate in a southerly direction suddenly reversed itself heading north (Stone et al, 1982). Secondly, it was during this period that Wrangellia collided with the Peninsular Terrane to form the Southern Alaska Superterrane (Csejtey et. al. 1979). Thirdly, in their present configuration in Southern Alaska, the Peninsular Terrane is thrust northward over Wrangellia (Coney and Jones 1985). And finally subduction in the Peninsular arc's trench suddenly ceased about 154 million years ago (Moore and Connelly, 1979) at about the same time as island arc volcanism resumed in Wrangellia 149-120 million years ago (Nokelberg et al. 1985). The above scenario attempts to account for the coincidence of these events.

4. Several authors date the initial collision of the Superterrane with North America as mid to late-Cretaceous but argue from different lines of evidence. They include: structural and stratigraphic evidence from the central Alaska Range (Jones et. al. 1982), a marked increase in sedimentation in what are now the upper Cook Inlet, Matanuska Valley and Copper River Basin areas suggesting the establishment of a forearc basin, the opening of the Border Ranges Trench on the trailing edge of the Superterrane at about 120±15 million years ago (Pavlis, 1982), the resetting of the atomic clocks of the Nikolai Greenstones 112 ± 8 million years ago (Silberman et. al. 1979), the emplacement of the East Susitna batholith 100-110 million years ago and mid-Jurassic and Cretaceous volcanism and plutonism in similarly dated accretionary wedge of Gravina-Nutzotin terrane (Nokleberg et al. 1985).

5. Estimates of the site of collision of the Superterrane with North America vary from the vicinity of present day California, off British Columbia, off Southeastern Alaska to near the present Wrangell Mountains in Southcentral Alaska. Assuming that the Superterrane approached from the southwest, the present position of the southernmost fragment of Wrangellia which forms the common Oregon, Washington, Idaho border suggests this as the most likely southern limit for this collision. The collision zone probably stretched for many hundreds of miles from this location. Our scenario assumes this location as it places the Superterrane in the right position for the formation of the rocks of the Prince William Sound region — rocks that seem to be derived from the Wrangellia and Peninsular terranes (Nelson et al. 1985, Dumoulin, 1987) and perhaps partially from continental sources (Moore et. al. 1983, Plafker, 1988). Secondly, a Jurassic micro-fossil found in the McHugh melange suggests affinities with Oregon (Nelson et al. 1985) and a Cretaceous Ammonite found at Ziegler Cove is one commonly found in the Pacific Northwest (this study). Coe et. al. 1985 in reviewing the accumulated paleomagnetic data for Southern Alaska suggest the latitude of Cape Mendocino in Northern California which is in good agreement with the above.

A recent paleomagnetic study of the rocks now in the northern fragment of Wrangellia suggests that they have moved from this southerly latitude since latest-Cretacious times (Panuska, 1985). Although this paleomagnetic study places these rocks farther to the south (at 32°) than present day, northern Oregon, we must remember that in the past 100 million years this portion of the North American Continent has undergone considerable westerly and slight northerly drift due to a counterclockwise rotation. If we walk the North American continent back along its line of drift, we discover that 90 million years ago the Hells Canyon area of Oregon was at the present latitude of San Diego (32°) and 60 million years ago just north of present-day San Francisco (40°) which is in good agreement with paleomagnetic data obtained for the Chugach/Prince William Terrane (Plumely et al. 1983).

6. There is little agreement on the final emplacement of the Superterrane, Chugach, and Prince William terranes in Southern Alaska. For example, on the basis of the Jurassic-Cretaceous flysch, Csejtey and others (1982) assign a Cretaceous date to this event. Hillhouse et al. (1985) showed that the Alaskan fragment of Wrangellia had to be in place before 50 million years ago. Harbert (1987) argues that the docking of the Peninsular Terrane ocurred before 45 million years ago. Keller and others (1984) and von Huene and others (1985) argue that The Prince William Terrane was in place by 42 million years ago. Finally Cowan indicates a displacement of over 650 miles for the Chugach Terrane after 40 million years ago. If our scenario above is correct, we may have a classic case of "blind men observing an elephant." One can not necessarily generalize from the part to the whole. Telescoping, fragmentation and intermittant and halting progress of the northward moving terranes probably created a series of ever changing relationships and perhaps a number of different arrival times. Secondly, not knowing the longitudinal extent of these terranes either as a whole or individually, it is difficult to generalize a measurement made at a single geographic point to the Superterrane as a whole. Hence, one might expect a certain scatter in estimates for the time of final docking in Southern Alaska for the Superterrane.

7. That the Border Ranges Trench has a separate identity from the Aleutian Trench is suggested by more than just paleomagnetic studies of their relative locations throughout the Mesozoic and Cenozoic eras. Although the ancient Koryak Trench may have joined an older Mesozoic trench that bordered the entire west coast of North America, geological evidence, present terrane geometry and plate reconstructions suggest that during Cenozoic times this trench along southeast Alaska and British Columbia was transformed into a strike-slip margin against which the Superterrane made its journey north. Hence, evidence of this trench must be sought inboard of the Superterrne, whereas the fossilized Border Ranges Trench is located outboard of the Superterrane but inboard of the Chugach-Prince William Terrane. The interpretation above suggests that because of oblique subduction of the Kula Plate the Border Ranges Trench on the outboard edge of the strike-slipping microplate remained active into Eocene times while the older, Mesozoic trench was being transformed through oblique underthrusting and strike-slip motion into the coastal range metamorphic belt of Southeast Alaska and British Columbia. This metamorphic belt in places is bordered by thin arcs of deformed Jurassic-Cretaceous flysch presumably manufactured in this older, Mesozoic trench or was carried northward with the strike-slipping Superterrane.

8. In addition to being apparently younger than the inboard subterranes which have been extensively intruded by basalts, the sandstones of the Sitkalidak formation and the Montague belt are remarkably similar (Dumoulin, 1987), and both of these groups exhibit landward structural vergence (Byrne and Hibbard,1987). Whether this landward vergence is a result of an increased geothermal gradient associated with the renewal of the Aleutian Trench or the result of a ridge-trench interaction is a matter for speculation. (C.f. Byrne and Hibbard, 1987 for a discussion of the influence of elevated geothermal gradients on the formation of landward vergence in an accretionary wedge).

9. For a more complete treatment of the glacial history of Prince William Sound see our companian volume — Nancy R. Lethcoe, *An observer's guide to the glaciers of Prince William Sound* (1987).

Chapter 3. Observing Prince William Sound's Rocks

Introduction

Fortunately, there are not a great variety of rock types in Prince William Sound and the novice with a little practice can become fairly adept at recognizing most of the common varieties. Geologists classify rocks into three general catagories depending upon their mode of origin. *Igneous rock* is rock which has been crystallized from a molten magma. If the rock has been extruded beneath the oceans or beneath the atmosphere to cool quickly into a fine crystalline form, it is known as a *"volcanic"* or a *"lava."* If it intrudes the crust and cools slowly into a coarse crystalline form, it is known as an "intrusive" or as a "plutonic" rock. *Metamorphic rock* is rock that has been recrystallized in a solid state under the influence of heat and/or pressure. Any of the three major rock types may undergo metamorphism. *Sedimentary rock* is rock formed usually in an aqueous environment from the fragments of other rocks (clasts). The fragments can vary from extremely fine (siltstones) to extremely coarse (conglomerates) and are often cemented together by finer fragments known the matrix. The cementing process occurs under pressure when the sediments are compacted and is known as *lithification.*

Over long periods of time, each rock group can be transformed by geologic forces into each of the other groups. This process ocurring over millions upon millions of years is known as the rock cycle.

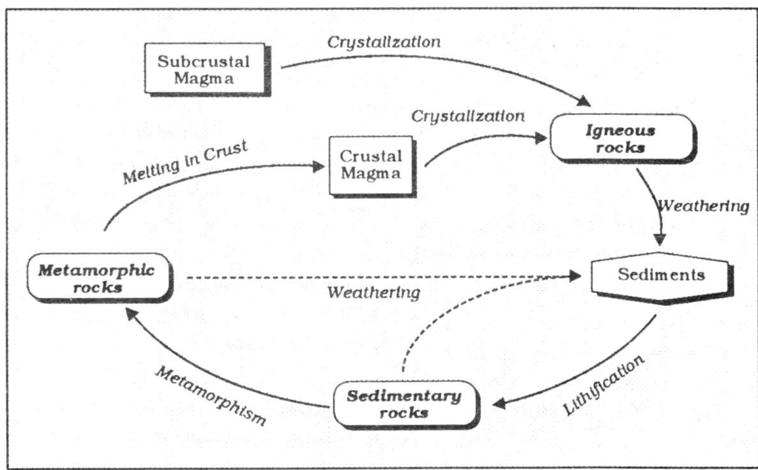

Fig-45. Rock cycle.

Observing Prince William Sound's Volcanic Rocks

Volcanic rocks in Prince William Sound consist of basalts that were extruded onto the ocean floor either at a mid-ocean ridge or through a leaky transform fault. These lavas differ from land-based flood basalts as contact of these heated rocks and magma with the minerals in seawater have chemically altered them. Notably they have become enriched in sodium and somewhat depleted in calcium, magnesium, and silica. The released calcium and magnesium ions can combine with dissolved carbon dioxide in seawater to form into magnesium enriched limestones similar to those found at Ellamar while the released silica is often lithified into chert. Small deposits of limestone and chert are sometimes associated with the basalt flows of the Sound. Secondly, the minerals epidote and chlorite form as a by-product of this reaction giving the lavas of the sound a slightly greenish tint. Regional metamorphism of basalt also results in the creation of epidote and chlorite and may have likewise contributed to the greenish color. For this reason the basalts of the Sound are called "greenstones." Greenstones are heavy, dark, mafic rocks usually exhibiting a fine crystalline structure caused by rapid cooling of lava on contact with cold water. In Prince William Sound greenstones assume three distinct forms: pillow basalts, sheeted dikes, and massive flows.

Pillow Basalts

Pillow basalts form when magma derived from the earth's mantle spills out onto an inclined surface of the sea floor. Upon contact with the cold water, a glassy surface layer forms around a molten core which still continues to flow. The confining pressure of the surrounding water molds this plastic mass into a roughly spherical shape as the lava oozes forth like toothpaste. Sometimes the molten core of an elongate pillow will burst the surface forming subsidiary multi-lobed pillows.

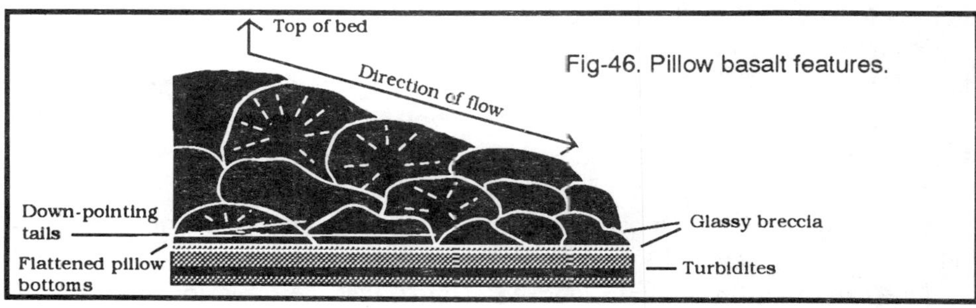

Fig-47. Pillow basalts. Note how overlying pillows conform to the tops of underlying pillows and the presence of downpointing tails. Photo curtesy of S. W. Nelson, USGS.

Because of its molten core, the basalt is still somewhat plastic when it comes to rest; the bottoms of the pillows readily conform to the underlying surface. Pillows flowing over older, chilled pillows often overlap two or more of these. When this happens, gravity tugs at overlying plastic mass causing it to fill the roughly triangular spaces between pillows. This "tail" (or "keel") points to the bottom of the bed. When the lava flows out onto a relatively flat seafloor covered with sediments, the bottom of the pillows are flattened indicating the bottom of the bed. Pillows are often elongated in the direction of flow. Sometimes, the contact between hot magma and cold water is so violent that the surface of the pillow explodes producing a burst of glassy shards which forms a glassy breccia filling the interfaces between pillows. Under some circumstances, one finds whole flows where entire pillows have been shattered by this violent interaction. Geologists refer to these shattered pillows as "pillow breccia."

One can often find pillows cracked open so that their interiors are exposed.

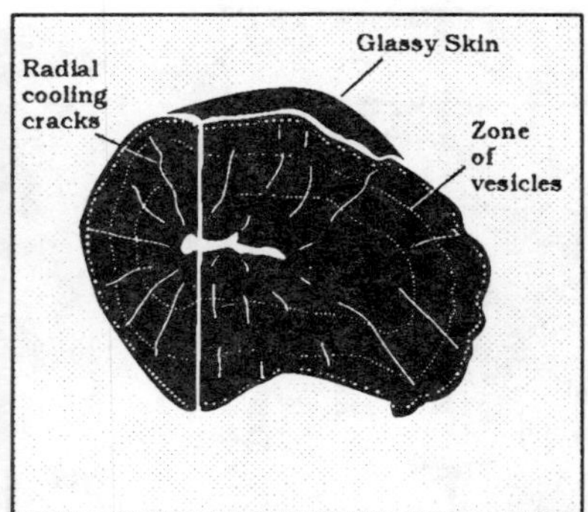

Fig-48. Left. Cracked pillow basalt showing zone of vesicles, radial cooling cracks, and breccia. Courtesy of S. W. Nelson, USGS. Fig-49. Above. Anatomy of a pillow basalt. After Compton (1962).

Here, a glassy surface gives way to a fine, crystalline interior. Usually, radial cracks resulting from contraction on cooling are visible. Sometimes, one can observe rings of tiny pockets filled with a white or greenish mineral; these represent the former sites of tiny gas bubbles called "vesicles."

Sheeted Dikes

Sheeted dikes represent the feeder vents that supplied the extruded pillow basalts. These most likely had as a source a shallow magma chamber near an oceanic spreading center. Repeated intrusion of older dikes and pillows by progressively younger dikes creates sheets of vertical intrusions resembling a deck of cards standing on end. On central Knight Island and on Glacier Island the cliffs of sheeted dikes exhibit distinctive, almost columnar, vertical structures. Seawater seeping down through cracks in the seafloor have altered these basalts to greenstones similar to those of the overlying pillows. The alteration to greenstones may also be due to regional metamorphism during uplift.

Fig-50. Sheeted dikes crosscutting Prince William Sound massive basalts. Photo by S. W. Nelson, USGS.

Massive Flows

Relatively featureless masses of basalt which lack pillow or vertical dike forms are referred to as "massive flows." These may represent large, lakelike extrusions of basalt onto a flat ocean floor where there was insufficient flow for pillows to form, or they may represent intruded sills.

The greenstone lavas of Prince William Sound are found overlying slate and graywacke beds (Galena Bay), lying beneath conglomerate beds (Galena Bay), and intruded into slate and graywacke beds (Evans Island, Knight Island). The interlayering of greenstone lavas and turbidites imply that these lavas erupted repeatedly and were repeatedly covered by turbidity flows or that they intruded preexisting trench-fill sediments.

Fig-51. Pillow basalts intruding into Prince William Sound turbidites at Squire Island anchorage.

This association of greenstone flows with trench-fill turbidites in Prince William Sound suggests that these rocks were formed by a spreading center interacting with a subduction trench.

Ophiolites

Knight and Glacier Islands in Prince William Sound represent cross sections of ocean crust emplaced on land and are called "ophiolites." On both islands one can see sheeted dikes underlying the pillow basalts they fed during the period of submarine volcanism. Thes slivers of ocean crust probably represent young ocean crust, probably sections of the Kula-Farallon Ridge, that were too warm and bouyant to be subducted in the Border Ranges Trench. Hence, they were probably thrust across the trench to become incorporated into the crust of the Alaskan continental margin.

Fig-52. The Knight Island Ophiolite. The columnlike structures of the mountains in the background are sheeted dikes.

Fig-53. Glacial erosion has exposed the core of a mountain where the granite pluton (lighter gray) has intruded the overlying sedimentary rocks near Nellie Juan Glacier.

Observing Plutonic Rocks

Plutonic rocks are composed of minerals which have been crystallized from a partial melt deep in the Earth's crust. As the molten rock rises toward the surface, it cross-cuts and intrudes the preexisting sedimentary beds finally cooling into a solid granitic mass. These rocks, later uplifted and eroded by glaciers, reveal to the trained eye subtle records of these ancient events.

Rocks that have crystallized slowly from a confined melt typically exhibit a mosaic of large, interlocking crystals visible to the naked eye. Often larger crystals will be observed set in a finer crystalline ground-mass suggesting that the larger crystals may have grown slowly at depth while the finer matrix formed as the molten mass cooled more rapidly closer to the surface. These rocks are usually light gray or pinkish in color and have a characteristically speckled appearance due to the presence of the dark, "mafic" (iron-magnesium) compound — hornblende and biotite mica. The predominating lighter crystals are the "felsic" minerals — feldspar and silica (quartz). Although popularly referred to as "granite," geologist distinguish a number granitic intrusives depending upon the kinds of feldspar, the amount of quartz and the amount and kinds of darker mafic minerals present.

A single pluton may exhibit a number of these different intrusives depending

Fig-54.	INTRUSIVES		
Felsic	Intermediate		Mafic
← Increasing silica (lighter color) ———————————————			
Granite	Granodiorite	Diorite	Gabbro
—————— Increasing iron and magnesium (Darker speckles) —————→			

upon its particular cooling history. The crystals in a cooling magma chamber may sort themselves out in a variety of ways creating rocks of slightly different compositions. Because the felsic minerals have a lower melting point than the mafic minerals, they may remain in the melt longer thus enriching its composition in these minerals near the end of the cooling cycle. In this manner a more felsic, granitic rock may be fractionated from a mafic, gabbro melt. It may be that the small, granodiorite bodies found in the mafic Knight Island greenstones originated in a such a manner. Secondly, crystals of differing densities may exhibit different settling rates in the still viscous melt causing different minerals to settle out into distinctive layers. Indeed, many plutons exhibit banded structures of segregated layers of differing mineral compositions. Finally, different ions in the melt exhibit differing rates of thermal diffusion toward the cooling margins leading to segregated halos of minerals of differing compositions.

As a plutonic mass cools, its outer rind chills more rapidly than its inner core, hence one might expect to find larger crystals in the interior of a pluton than at its "chilled margins." Although this is often the case, convection currents in the crystalline mush may distribute temperatures more evenly creating a more uniform distribution of crystal sizes. However, these currents themselves often leave clues as they flow past the walls of the magma chamber. Here at the interface between the hot magma and the cold country rock, one can often see the darker, elongated mafic crystals of the granite aligned parallel to each other and what must have been the former chamber wall. If the magma was hot enough to cause the sedimentary rock to flow plastically, wisps of sedimentary material may be caught up in the magmatic flow. These features are called "flow structures."

Fig-55. Flow structure. This turbidite rock from the magma chamber wall was melted and swept up in the flowing magma. Photo courtesy of S. W. Nelson, USGS.

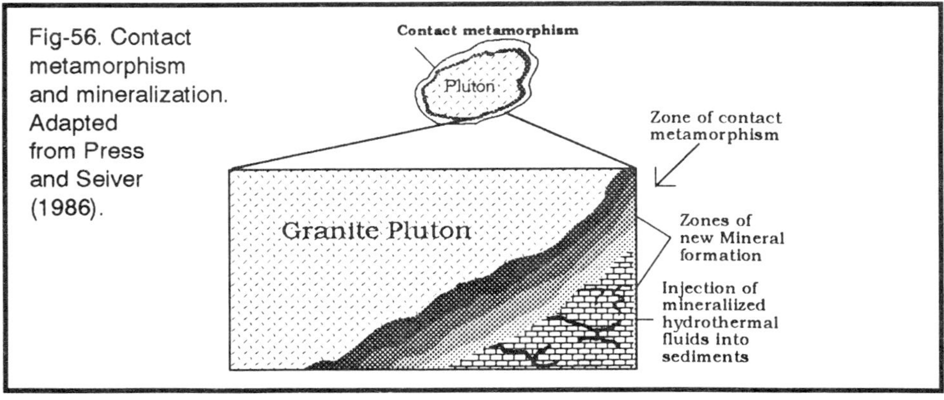

Fig-56. Contact metamorphism and mineralization. Adapted from Press and Seiver (1986).

The interface between the magma body and the sedimentary country rock is called the "contact zone" and should be examined closely; for it records events surrounding the emplacement of a particular pluton. This area of baked sedimentary rock surrounding the granite may extend a mile or more outward from the pluton. The grade of contact metamorphism decreases with increasing distance from the magma body. The intensely baked sedimentary rock at the immediate edge of the contact zone is often dark, dense and of a fine, random crystalline structure. This contact metamorphic rock is known as a "hornfels."

Chunks of metamorphosed sedimentary rock are often found "floating" in the solidified crystal melt near the contact zone. These are called "inclusions." Inclusions of country rock can be quite large. In the Sheep Bay Pluton, for example,

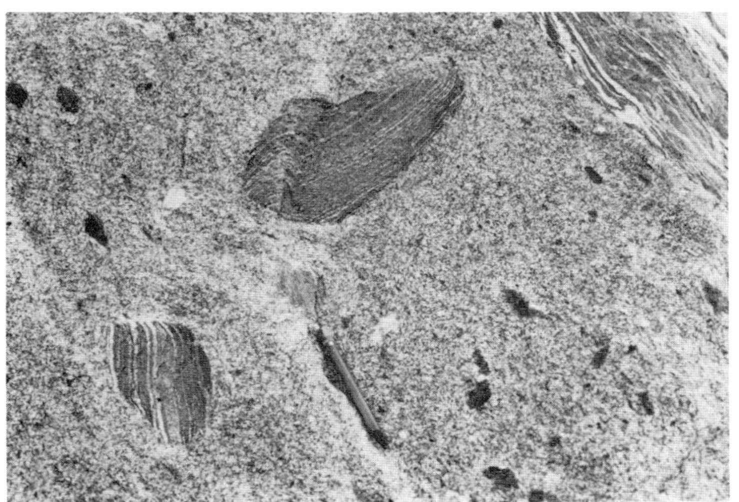

Fig-57. Inclusions. Chunks of metamorphosed sedimentary rock are often found "floating" in the solidified crystal melt near the contact zone. Note the foliated contact zone in the upper right hand corner. Photo courtesy of M. Miller, USGS.

a 300 foot in diameter section of the magma chamber roof seems to have collapsed into the magma pool and now appears as a giant inclusion.

If the pluton has been forcibly emplaced, then the metamorphosed country rock may show signs of foliation indicating that the magma body probably not only metamorphosed the country rock thermally but also exerted considerable pressure. Geologists believe that one mechanism that allows a pluton to rise through country rock is that it heats the rock until it flows plastically then wedges it aside. The hot magma may also intrude the country rock to such an extent that large sections of the magma chamber's walls and roof fall as large inclusions into the molten pool to become later incorporated into the melt itself. If the contact zone has been subjected only to thermal metamorphism, original sedimentary bedding and even cross bedding may be distinguished.

The sharpness of the contact provides geologists with information concerning the relative depths at which a pluton may have cooled. A gradational contact in which the zone of mixing of the intrusive and intruded rocks extends over a broad area suggests slow cooling during deep burial. Most of the plutons in Prince William Sound exhibit sharp contacts with the country rock and tend to cross-cut the sedimentary bedding planes suggesting that they may have cooled at relatively shallow depths of less than 4 miles.

Dikes and sills represent minor injections of differentiates of the magma body into the country rock. Dikes are discordant or oblique to the bedding planes cross-cutting them; while sills represent molten material which has been injected between bedding planes.

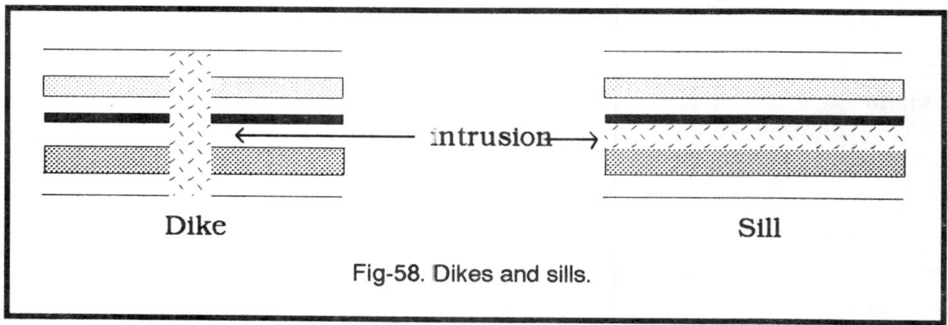

Fig-58. Dikes and sills.

Feldspar and quartz are common dike minerals which sometimes carry with them minerals of economic importance such as gold, silver, zinc and lead. Dikes and sills, originating from molten rock, exhibit flow structures, inclusions and chilled margins; but because they are smaller bodies and cool more rapidly, they in general have a finer crystalline structure than their parent plutons. Examining

Fig-59. Billings Glacier Pegmatite. Giant crystals form as dikes within the surrounding granite when groundwater currents move ions through the melt. Photo by S. W. Nelson, courtesy USGS.

the contact zone for dilation of the wall rocks may reveal whether a dike was forcibly injected at depth or merely filled a preexisting fracture in the country rock at shallower depths where these rocks are more brittle.

Ground water circulating around the cooling pluton may play a great role in determining the differentiates which are injected into dikes. Water in a melt makes it less viscous allowing it to flow more freely; consequently, melts exposed to circulating groundwater may be more influenced by internal convection currents which move ions more efficiently through the melt thus promoting rapid crystal growth. Under such conditions, giant crystals may form as dikes within the granite or the surrounding country rock. These are known as "pegmitites" and often contain excellent mineral specimens.

When Plutons contract during cooling, perpendicular sets of cooling joints tend to form so that granite exposed to freeze-thaw conditions often fractures into great blocks. Exposed granite faces also often fracture off in great sheets. This "sheeting" is thought to be the result of off-loading. Plutons form deep within the crust under considerable pressure caused by overlying rocks. When this overburden is removed by glacial erosion, the granite is under considerable tension and may spontaneously explode sloughing off great sheets (cf. Fig-60).

Because of their chemical compositions, plutons are easily dated radiometricaly using the potassium-argon method. The Sheep Bay Pluton in the southeast corner of the Sound has been dated to be on the order of 52 million years old. Since this pluton cuts the bedding planes of the sedimentary Orca rocks of this region, geologists are able to determine that these sedimentary rocks have to be at least of

Fig-60. Sheeting may occur when glaciers remove overlying sediments releaseing pressure on the buried granite which may spontaneously explode sloughing off great sheets as at Greystone Bay, Port Nellie Juan.

Eocene age. This and other Eocene plutons stretching from southeast Alaska to Kodiak seem to be a part of a belt of plutons thought to have been related to the subduction of the Kula-Farallon ridge (Moore, 1983).

Intruding the Valdez rocks of the northwestern corner of the Sound between Port Wells and Eshamy Bay is a group of plutons dated to be about 35 million years old (Oligocene). These plutons are characterized by a younger, mafic, gabbro phase.

Observing Metamorphic Rocks

Metamorphic rocks are sedimentary, igneous, or previously metamorphosed rocks which have been recrystallized by heat and/or pressure while still in a solid state. Except for the addition or subtraction of water and carbon dioxide, the minerals of most metamorphic rocks retain their original chemical identities but assume new crystalline forms. These new crystal structures grow under unique circumstances of heat and/or pressure giving geologists clues as to the conditions that formed them.

Rocks buried deep within the crust experience increased pressure and heat which lead to recrystallization. Likewise, sediments carried down into a subduction trench in a sedimentary prism are buried, compressed and folded, and often scrapped off onto the underside of the continental plate (underplated) undergoing

Fig-61. Foliation occurs when sedimentary rocks are subject- ed to both heat and pressure which gives them a wavy, banded texture and sharp cleavages.

lithification and metamorphism. Since most of the sedimentary rocks of Prince William Sound were formed in this manner, most are metamorphosed to some extent. Hence, some geologists prefer to classify the Sound's turbidites as "metasediments." The earthquake wrenching forces of uplift which have warped the Chugach Range into its present contorted form have further folded, faulted and metamorphosed much of the Sound's rock. Where hot, granitic plutons have intruded the sediments, the rocks have been baked and recrystallized - sometimes for a mile or more surrounding the intrusion. Likewise, smaller metamorphic zones, measured in inches, may be traced along felsic and mafic sills and dikes injected into the sedimentary rock. Where faults have rent the Sound's surface, the rocks have been stressed, fractured and ground into a fine powder. These frag- ments, referred to as "gouge," have often been recemented into a fault breccia whose recrystalized, mineral structure differs from that of the surrounding rock.

Metamorphic rocks which have been subjected to both heat and pressure are the most easy to recognize for they often exhibit a wavy, banded texture and often cleave, cleanly along surfaces oblique to the original bedding planes. This leafy, texture is referred to as "foliation" because of its resemblance to foliage.

The white banding is often due to the linear formation of light colored feldspar and quartz crystals. Such rocks because they have been metamorphosed during subduction or uplift are of a regional extent and are referred to as "regionally metamorphosed."

On closer inspection, one can see that the crystals are solidly intergrown and some are aligned in a single direction giving the whole rock a streaked or banded appearance. Thus, aligned often elongated crystal structures provide an important

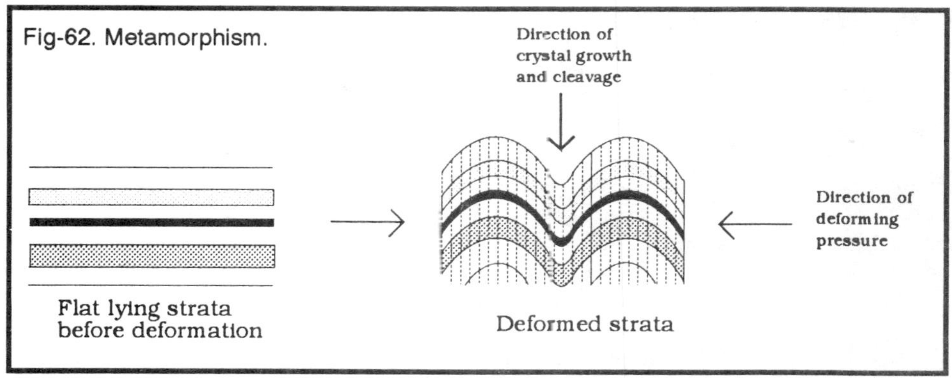

clue as to the identification of metamorphic rocks. The linear arrangement of crystals is a result of the directed pressure on the rock; crystals tend to grow along paths of least resistance and hence perpendicular to the applied pressure.

This linear crystal growth also explains why metamorphic rocks formed under heat and pressure tend to exhibit well defined cleavage. Flat, platey crystals of certain minerals (especially mica) become aligned in the rock during recrystallization. When the rock is struck a blow, these crystals tend to slide easily past one another causing the rock to fracture in a single plane.

The following photo (Fig-63) illustrates how pressure distorts the fabric of metamorphic rock. The tiny quartz vein most likely originally cut across the fine bedding (laminae) of the sedimentary rock in more or less a straight line. When the rock was subjected to pressure parallel to these bedding planes (from the left and right sides of the photograph), they slipped offsetting and contorting the intruded quartz vein.

Highly metamorposed beds sometimes exhibit a sausagelike structure where a relatively competent layer caught between two, more plastic layers has been stretched and thinned until it ruptures into elongated, discontinuous, sausagelike cylinders called "boudins." Such metamorphic rocks are described as "boudinaged."

Rock metamorphosed by heat alone differ somewhat from the regionally metamorphosed rock described above. The crystals formed when a rock has been baked by proximity to a pluton tend to be fine-grained and exhibit a random rather than linear arrangement. These usually dark, tough, fine-grained, non-foliated rocks are called "hornfels" and are most easily identified by their location in the contact zone.

These baked sedimentary rocks range from preservation of to complete destruction of original bedding features. Less intensely baked rocks farther from

Fig-63. Note the highly distorted quartz vein above the pencil. Pressure may distort quartz veins causing originally straight veins, which cut across bedding planes, to become offset and contorted. Photo courtesy of M. Miller, USGS.

the contact often have a spotted appearance. Rocks recrystallized in this fashion are said to be the product of "contact metamorphism."

Geologists refer to metamorphic rocks as being of a low, medium or high grade depending upon the intensity of the heat and/or pressure which formed them. Certain minerals exhibit unique crystal structures under differing conditions of heat and pressure and can be used to identify the metamorphic grade of a particular rock. Most recrystallized rocks in Prince William Sound are of a low to medium grade. By far the most common, low grade metamorphic rocks in the Sound are the familiar black slates which are easily recognized by their conspicuous cleavage. Slate is the metamorphosed form of the claystone – shale. Whereas the softer shale tends to fractures parallel to its bedding planes, growth of mica crystals in the rock in response to heat and pressure cause the harder slate to fracture obliquely to the bedding plane. If the shale is subjected to further heat and pressure the mica crystals grow in size and some are converted to feldspar. The resulting rock becomes harder and loses its cleavage so that it shatters when struck with a hammer. This dark, hard mudstone resembling slate but lacking its cleavage is known as "argillite." Further conversion of larger quantities of mica to feldspar gives the rock a slightly banded, foliated appearance. The shale has become a medium grade, metamorphic rock called a "schist." If enough mica crystals remain, schists may continue to exhibit some cleavage. If this schist is further metamorphosed, it becomes a high grade gneiss (pronounced "nice"). The conversion of over half the mica to feldspar gives the rock a definite banded appearance and it loses its cleavage. The metamorphic grades of rocks likely to be encountered in Prince William Sound are summarized in the following table.

Fig-64.

Metamorphosed Sedimentary Rocks of Prince William Sound

Increasing metamorphic grade →

Type	Grain Size	Foliation	Cleavage	Occurance in PWS
Clay	Fine	NA	NA	Common
Shale	Fine	None	Fractures Parallel to Bedding	Common
Argillite	Fine	None	Shatters	Fairly Common Locally
Slate	Fine	None	Oblique To bedding. Cleaved Surface has Dull sheen	Common
Phyllite	Fine but Coarser than Shale	Some Often Crenulated	Oblique, Wavey. Sparkling Surfaces (mica).	More Common in Valdez group.
Schist	Coarse Visible Mica	Thickly Foliated	Oblique to Bedding	Some Locations
Gneiss*	Medium to Coarse	Thinly Foliated. Banded & Streaked	None (mica Changed to feldspar.)	Rare but Present at Some Locations

* Not always more metamorphosed than schist

Observing Sedimentary Rocks

Sedimentary rocks are formed by the consolidation of sediments which have in most cases settled from an aqueous medium. Geologists classify sedimentary rocks according to their grain size and mineralogy. Fine-grained rocks with grain sizes of less than .0025 inches are called "mudstones." Rocks with a grain size of between .0025 and .08 inches are referred to "sandstones" while rocks which contain grains ("clasts") greater than .08 inches are called "conglomerates." Conglomerate clasts may vary from granule to pebble size and even to boulder size.

The minerals composing the mudstones, sandstones and conglomerates of Prince William Sound indicate that they were derived from the erosion of a volcanic arc eroded to its plutonic roots. Some geologists believe that the island arc terrane of the Alaska Peninsula may be the source of these sediments (Dumoulin, 1987). In the case of the Orca rocks, some geologists indicate that the Valdez rocks could be a likely source (Gergen and Plafker, 1988.)

Turbidites

Most sedimentary rocks in Prince William Sound are "turbidites" (sometimes referred to as "flysch"). Indications are that the original sediments were deposited by turbidity currents in a subduction trench (probably the Border Ranges Trench) roughly between 80 and 50 million years ago. Turbidity currents, often referred to as "density currents" or "gravity flows," occur when great quantities of water-logged sediments perched on a steep submarine slope begin to slump and slide. In a seismically active trench environment, earthquakes are often thought to initiate the flow. As the liquefied sediments gain momentum, turbulence develops violently mixing grains of mud, sand and gravel. The swirling vortices in these currents are often energetic enough to suspend small boulders in the flow. The density of this flow then is much greater than that of the surrounding seawater, hence currents may develop enough momentum to carry the suspended sediments hundreds of miles offshore. The swirling, suspended sediments give these bottom-hugging currents great erosive power. It may be that the submarine canyons which dissect the inner trench walls have been scoured and enlarged by turbidity flows. As these fast-moving currents spread out onto the deep trench floor, they begin to lose some of their momentum and turbulence and the coarser boulders, gravels and sands begin to drop out, followed by less coarse sands, and finally by fine muddy deposits. Under these conditions, a graded bed often results with the coarser

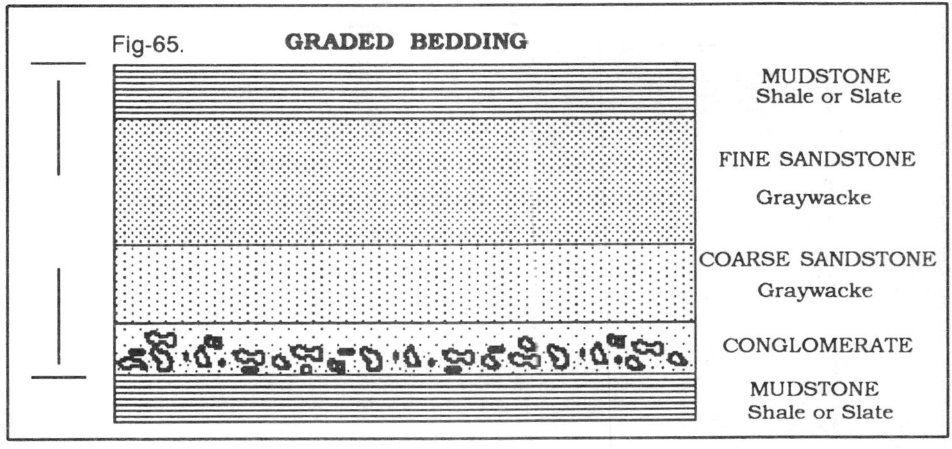

conglomerates lying at the bottom, grading upward to finer and finer sandstones and finally into a slate or shale mudstone layer.

In a normally graded bed, such as that depicted above, one can often distinguish the bottom of the bed from the top even though the bed has been severely tilted. First, the conglomerate tends to identify the bottom of the bed. Secondly, if one closely examines the contact between the two mudstone layers and the sandstone layers (assuming the conglomerate is absent as it often is), he will sometimes note a sharp contact in one instance and a graded contact in the other. The sharp contact represents the bottom of the bed where the heavier sand settled suddenly out of the current on top of the upper muddy layer of the previous turbidity flow. Because the finer sand, silt and clay settled out of suspension slowly (perhaps over a period of a week or so) the upper mudstone layers are finely graded. The sandstones of these beds are of a dirty, poorly sorted type called a "graywacke." The mixing in the turbulent vortices of the turbidity current probably accounts for the presence of fine-grained muddy, volcanic particles mixed in with the sand.

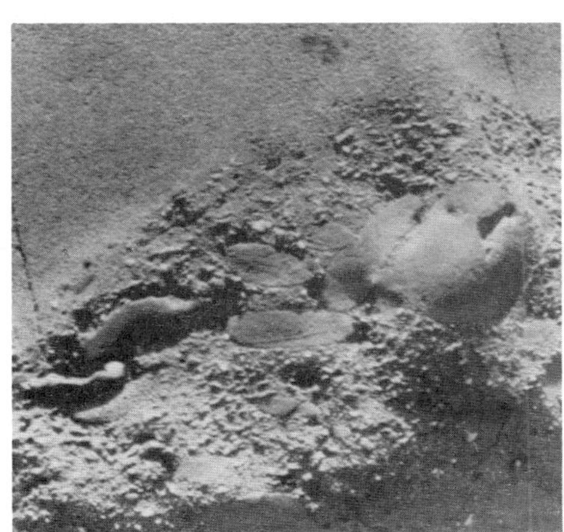
Fig-66. Bedded grading. Boulder, Green Island beach.

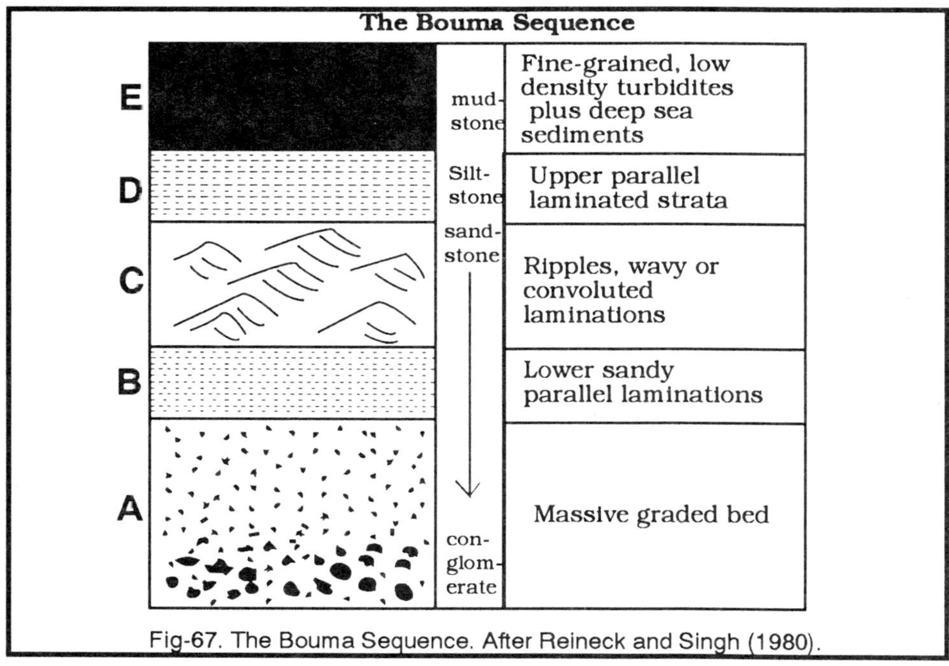

Fig-67. The Bouma Sequence. After Reineck and Singh (1980).

The Bouma Sequence

The graded bed depicted above (Fig-65) represents a simplification of what the geologist actually observes in the field. Actual, graded turbidite beds may reveal even further structures known as the Bouma Sequence. The bottom of the Bouma sequence (**A**) consists of a massive (i.e. no apparent bedding) sandstone/conglomerate layer grading up from a conglomerate, granulite or coarse sandstone to a medium grade sandstone. Above this, lies a region of medium to fine sandstone (**B**) revealing fine, horizontal beds (laminae) — perhaps indicating a shooting flow. Overlying this layer is a fine sandy to silty bed (**C**) characterized by ripple marks and convoluted

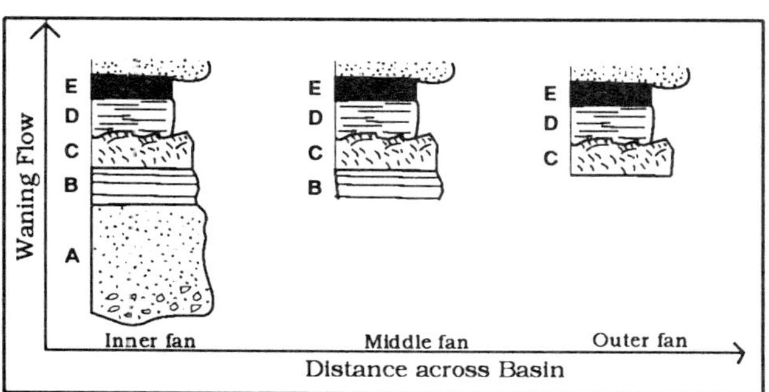

Fig-68. Buoma sequence showing waning flow toward the outer reaches of the fan. After Reineck and Singh (1980).

Fig-69. Top. Convoluted bedding flow typical of Bouma Layer C. Photo courtesy of S.W. Nelson, USGS.   Fig-70. Bottom. Ripple marks typical of Bouma Layer C. Green Island.

laminae — probably signaling turbulent flow. Layer **D** lying above this, consists of a silty layer of parallel laminae, indicating a probable return to a shooting flow. The upper, finely graded, muddy layer, **E**, represents the slow-settling, low-density components of the flow. If a considerable time elapses between flows, this layer may contain deep ocean sediments containing microfossils.

Complete Bouma sequences are rare in the field for the top layers may have been removed by erosion or the bottom layers may have already dropped from the current before reaching out farther on the turbidite fan.

Over millions and millions of years, successive flows accumulate on the trench floor forming a giant fan which is composed of a repititive sequence of graded beds. When these beds become lithified through compaction from the great overburden of sediments and accreted at a subduction trench, a repititous sequence of beds each composed of conglomerates underlying sandstones underlying mudstones (or sandstones underlying mudstones etc.) develops. Most of the rocks of Prince William Sound and indeed the entire Chugach range occur as repititious sequences of graded turbidite beds.

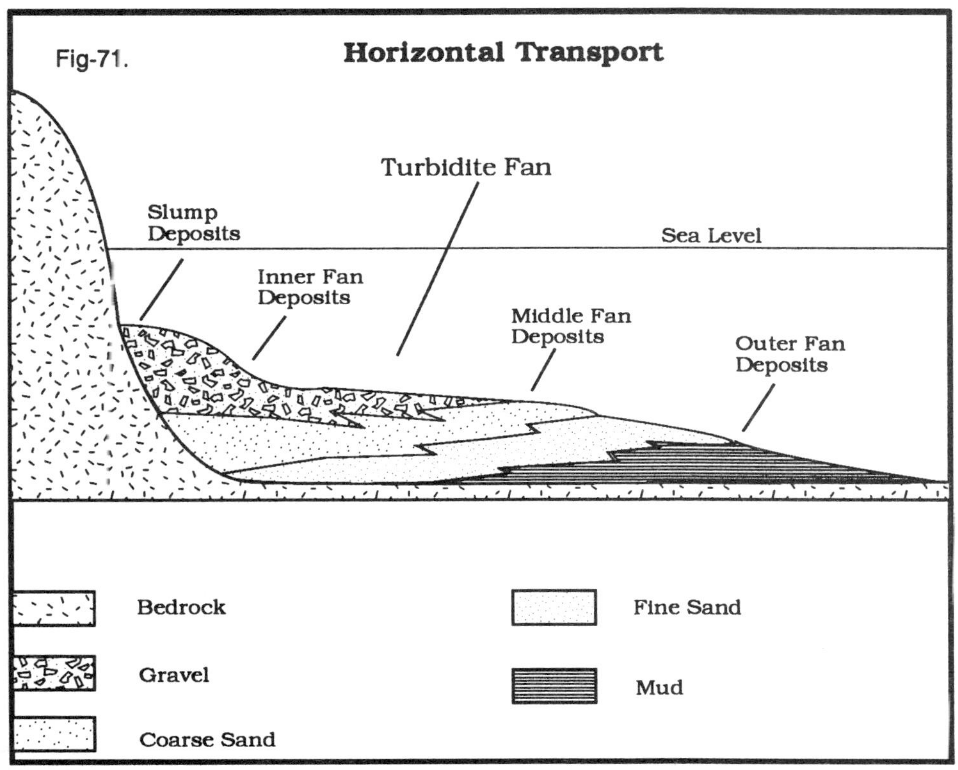

As each turbidity flow spreads out over the fan, it gradually loses energy and the heavier sediments drop out first, closer in toward the trench wall, followed farther out in the middle of the fan by less dense components, and finally by a blanket of fine muddy sediments on the farthest reaches of the fan. In this manner, the sediments are sorted horizontally as well as vertically across the fan.

Thus, a geologist studying a lithified cross section of a fan may be able to determine according to the Bouma sequence what part of the fan he is examining. Geologists have determined that the turbidites of Prince William Sound were deposited in an inner and middle-fan environment. Both complete and incomplete Bouma sequences can be found depending how far out on the fan each was deposited. Incomplete sequences often begin with division **B** or **C** (Winkler, 1976).

Conglomerates

The conglomerates of Prince William Sound tell a similar story of inner and middle fan deposition. Three general types of conglomerates are found in the Sound. In the first type the clasts seem to vary from being slightly angular to rounded and to be composed mostly of mudstone and sandstone pebbles often touching one another in a coarse sandy or granulite matrix. This kind of conglomerate tends to occur in very thick beds (measured in hundreds and even thousands of feet) and contains boulder-sized clasts.

The angularity of the clasts suggest a short erosional history while

Fig-72. Note the angularity of these clasts from the Rocky Pt. conglomerate beds in Galena Bay The clasts touch or nearly touch each other. Photo courtesy of M. Lewis.

the similarity in composition of matrix and clasts suggest a local depositional environment. The touching of many of the pebbly sandstone clasts imply that they dropped out of the turbulent flow before a great deal of mixing and sorting could occur. There are also relatively numerous gravel-size clasts to sand-size matrix clasts in these deposits. These indicators plus the presence of boulder-size clasts, which would be the first to drop from the flow, and the great thickness of the beds suggest that these conglomerates were deposited in the inner fan region — probably near the base of the trench slope.

A second kind of conglomerate, often found near the bottom of graded beds, is composed of clasts of well-rounded pebbles of quartz, granite, greenstones and other resistant rocks foreign to their present surroundings. Such "exotic" clasts may provide clues as to source terranes for the Sound's sediments. These clasts do not touch but are distributed throughout a medium to fine-grained, sandy matrix and are confined to beds measured in tens of feet in thickness. Most of the clasts average about an inch in diameter, boulder-size clasts are absent and the medium to fine sandy matrix is relatively more abundant than the conglomerate sized clasts.

The well-rounded and resistant nature of these clasts and their dissimilarity to the matrix and surrounding rocks suggest a longer erosional and transport history. The size of the clasts, thinner beds, and relatively greater abundance of matrix to pebbles suggest that these conglomerates were probably formed farther out on the fan. That the clasts are disbursed through a medium to fine-grained matrix rather than touching suggest that they were carried by the turbulent flow for a longer time than the first type of conglomerates. Geologists believe that these conglomerates may repre-

Fig-73. These conglomerates from Green Island consist of well-rounded pebbles which are distributed throughout a medium to fine-grained, sandy matrix suggesting a middle fan deposition.

Fig-74. Well-worn pebbles, distributed through a medium to fine-grained matrix, may indicate deposits laid down in distributory feeder channels that crisscrossed the middle part of a fan. Note the marked alignment indicating the direction of flow. McPherson Beach, Naked Island.

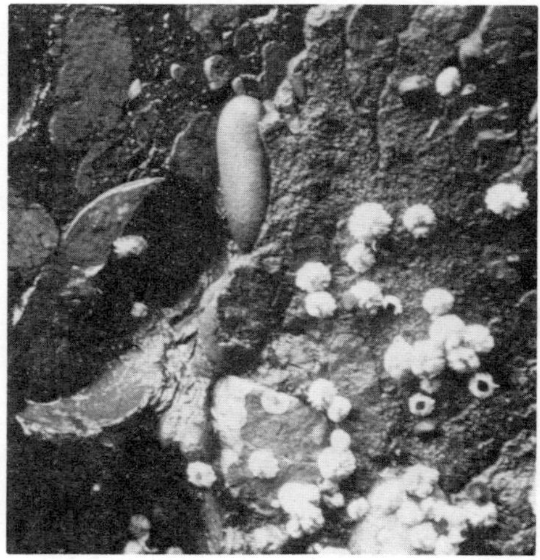

Fig-75. Sometimes conglomerates are founc which look like they have been slicec by an axe. These clasts and matrix are near a fault zone where they have been split by the force of an earthquake. Simpson Bay.

sent material transported in distributory feeder channels that crisscrossed the middle portion of the fan (Winkler, 1976).

Often the clasts in this type of conglomerate will exhibit marked alignment indicating the direction of current flow (Fig-74). Sometimes, if conglomerates have been deposited in an active fault zone such as those near the Bainbridge Fault, the clasts in the conglomerate may be sheared along with the sandstone matrix (Fig-75).

A third conglomerate type deposit consists of a poorly sorted, chaotic mixture of sandstone and mudstone containing very large boulders of the same materials. The massive and chaotic nature of these deposits and the absence of sorting suggest that they are not the result of turbidity currents so much as massive slumping of the inner trench wall and are referred to as "slump deposits." The Galena Bay conglomerate (Figs-76, 125) exhibits both slump and turbidity current components. The lighter, upper bedded layers suggest inner fan deposition by turbidity currents whereas the chaotic, clast supported, poorly sorted, boulder-filled, lower darker layer suggests at least partial slumping.

Fig-76. Angular and boulder sized clasts from the Galena Bay conglomerate bed showing inner fan deposition by turbidity currents.

Fig-77. Fragments of a consolidated muddy layer have been ripped up by a subsequent turbidity current and incorporated into the coarser-grained sandstone layer. Green I.

Rip-Up

A fourth kind of deposit in some ways resembling the above conglomerates is often found in the bottom sandstone layer of Prince William Sound turbidites. The clasts are angular pieces of mudstone often distorted into odd shapes embedded in a sandy matrix. Unlike the above conglomerate clasts which represent material transported over some distance, these deposits seem to represent the action of the turbulence on the ocean floor itself. If considerable time elapses between two turbidity flows, the upper muddy layer of the last flow may consolidate into a coherent layer. When the next turbidity flow occurs, the turbulent, erosive bottom sandy layer rips up fragments of this coherent, muddy bottom and carries them along with the current to deposit them downstream. When the sediments are lithified (compacted into rock) the ripped-up mudstone fragments appear as darker, fine grained inclusions in the lighter coarser grained sandstone (Fig-77).

One can see in the above photograph how fragments of the pliable mud have been bent and distorted by the current before deposition. Also, the parallel alignment of the elongate mud clasts indicate the direction of flow which was probably from left to right. Like conglomerates, rip-up can be used to establish the bottom of a turbidite bed.

Fig-78. Tilted turbidite beds. Erosion of the softer mudstone layer exposes the harder sandstone surface where sole markings can be found. Green Island.

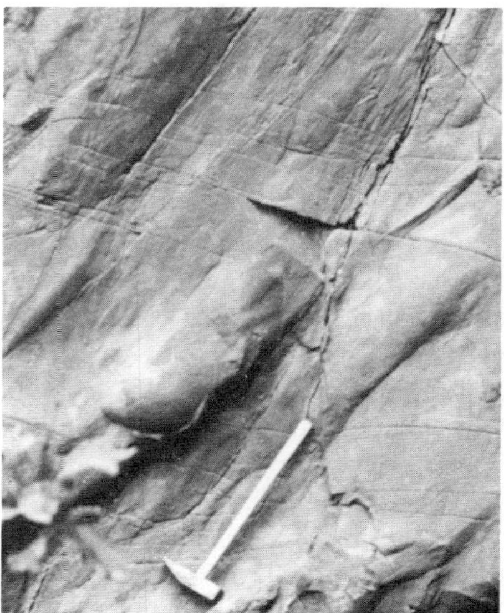

Fig-79. An exposed sandstone surface reveals both flute castings and groove marks. Photo by J. Dumoulin.

Sole Markings

The bottom, sandy layer reveals some of the most interesting features which can be observed in the Sound's abundant turbidite beds. Sand, gravel and other objects caught up in the bottom-hugging, turbidity current scour and erode the soft muddy layer of the previous flow which now forms the ocean floor. If the bottom is soft, we can observe a wide variety of erosional marks left in the wake of the passing current. However, we rarely observe these marks as scours in the mudstone but as casts in the bottom sandy layer. Mudstone is much less resistant to erosion than is sandstone; hence, when the mudstone is eroded away we observe the protrusions in the

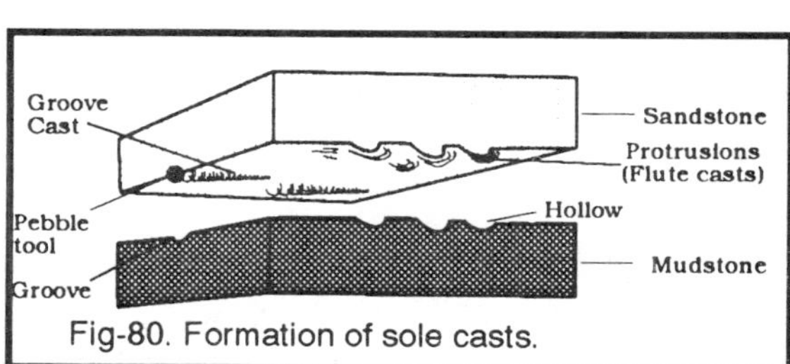

Fig-80. Formation of sole casts.

bottom of the sandstone bed which filled the hollows in the mudstone i.e. we see a cast.

Flute Casts

Flute casts are sole marks caused by the scouring action of grains of sand caught up in the vortices of a turbidity current. These swirling grains scour out a depression in the soft muddy layer which has a characteristic shape with a bulbous upstream end an a gradually flattening, trailing, downstream end.

As the current wanes the sand fills in the depression to become a cast on the bottom, sandy layer of the flow. Thus sole makings can be used to indicate the bottom of a bed and the direction of flow of the current. One geologist has studied a large number of casts on the eastern side of Prince William Sound and concluded that the rocks of this area were deposited on a westward sloping fan (Winkler, 1976).

Fig-81. Flute casting in tilted sandstone. The bulbous end marks the upstream, source of flow. Photo courtesy of S. W. Nelson, USGS.

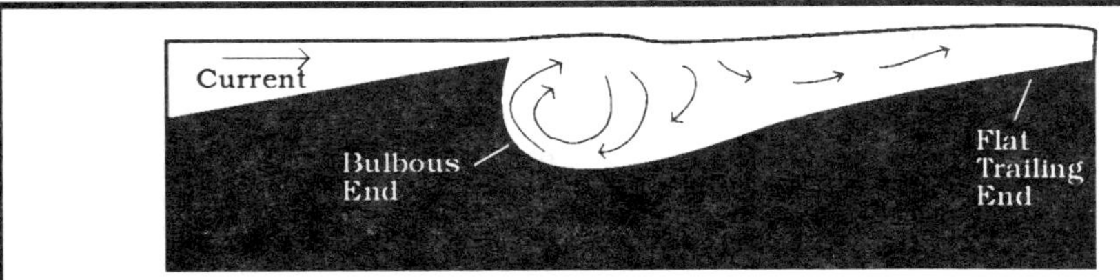

Fig-82. Flute casts result from turbidity current scouring of the mud bottom.

Tool marks

A wide variety of cast shapes have been made by larger tools carried by the current. These tools may consist of fish vertebrae, shells, bits of organic matter or pebbles which are dragged, rolled, or skipped across the bottom furrowing grooves or depressions in the soft mud. Later infilled with sand, these become casts. The shape of the object, the angle of incidence, the strength and turbulence of the current and the strength of cohesion of the muddy bottom may all influence the shape of these casts. The most common tool marks are "groove casts."

Groove Casts are relatively long, straight, gutter-like troughs which are caused by a current-propelled tool furrowing out a groove in a soft mud bottom. Occasionally, the tool responsible, often a pebble, is preserved at the end of a groove. However, more often the tool is lifted back into the current sometimes producing a sharp curve at the downstream end as the projectile is flung back into the current. In both cases one can determine the downstream direction of the current.

Most groove casts have clean, sharp edges but some shallow casts exhibit a series of v-shaped marks pointing downstream. These *chevron marks* are thought to be caused by a tool's just skimming the surface of a cohesive mud bottom. Turbulence caused by the passage of the projectile is thought to ruck up the surface into a series of chevrons. *Prod marks* are short, asymmetrical semi-conical or triangular depressions caused by a projectile striking the mud surface at a high angle of impact before bouncing back into the current. The asymmetry of the cast may be used to deduce current direction.

A lower angle of incidence produces a shallow, rounded, more or less symmetrical depression known as a *bounce mark.*

Brush marks like bounce marks are caused by projectiles having a low angle of incidence; but they exhibit a rounded ridge of mud where the tool exits the depression - possibly due to a different consistency of the mud surface.

Occasionally, tools will roll or skip across the mud bottom leaving a continuous or discontinuous series of sharp or rounded depressions known as *roll marks* and *skip marks*. These sometimes resemble glacial chatter marks.

Fig-83. Groove cast. Photo by Nelson, ISGS.

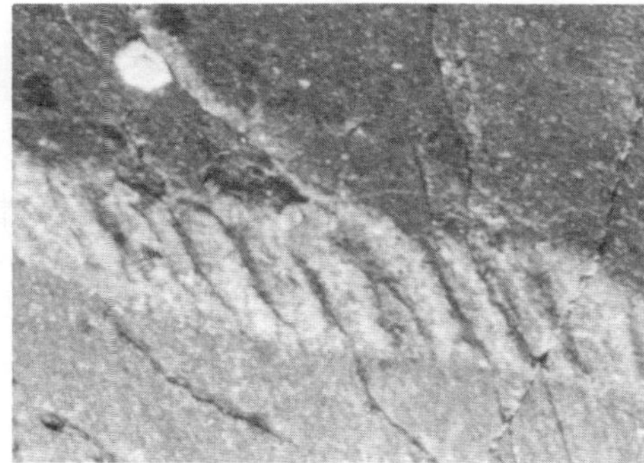

Fig-84. Chevron marks

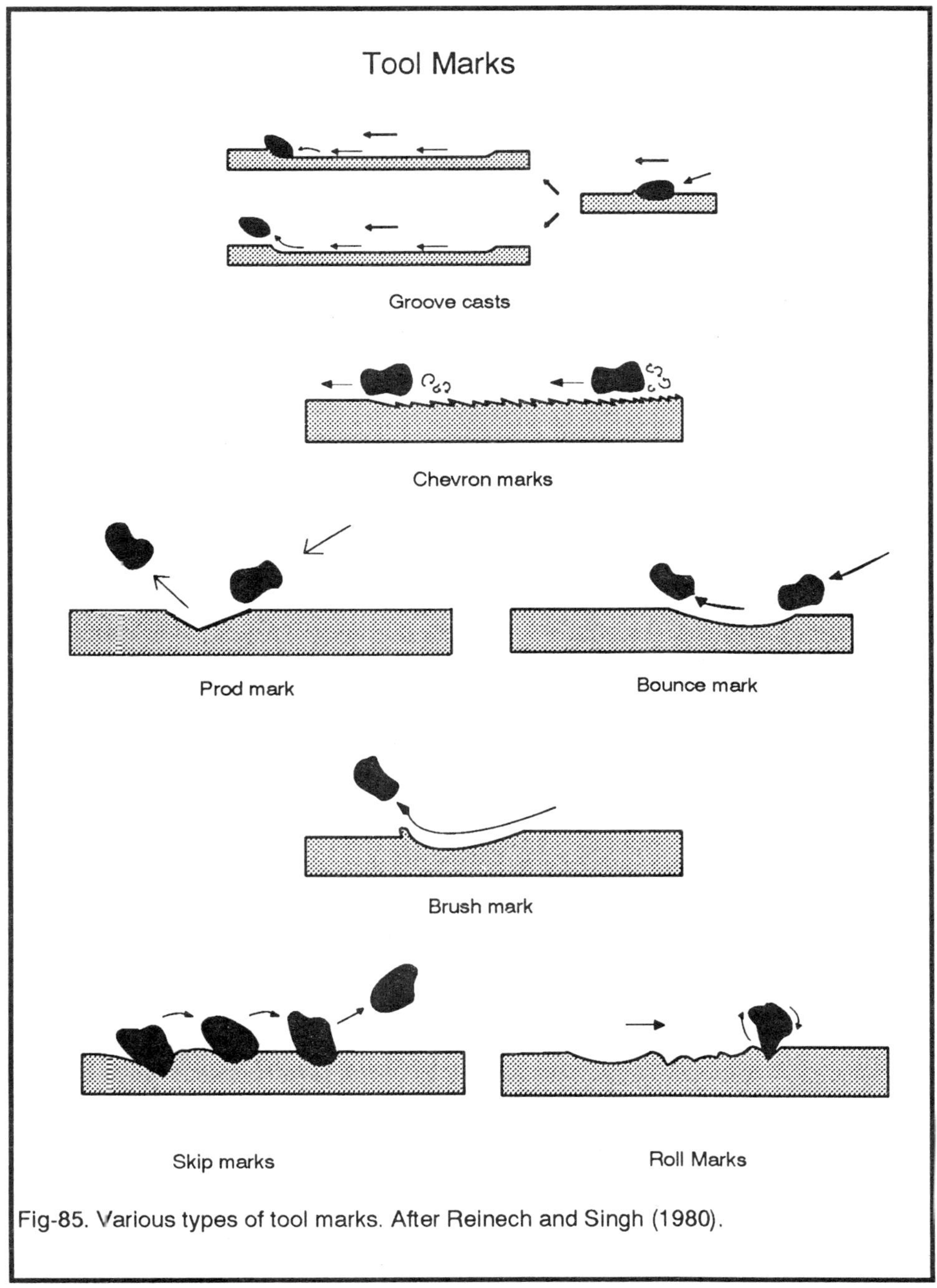

Fig-85. Various types of tool marks. After Reinech and Singh (1980).

Load Casts (Fig-86-87)

Load casts, unlike other sole marks, are formed after the passing of a turbidity current. Load casts are due to the difference in densities between the overlying sandy layer and the underlying muddy layer. For load casts to form, the underlying muddy layer must be sufficiently soft to allow the heavier sand to sink down into it. As the sand must displace the mud, it does not sink as a solid sheet but as numerous, discrete, bulges. The less dense, displaced mud then oozes up around the sinking, sandy knobs. When the mudstone erodes away one observes a load cast.

In certain places in Prince William Sound, one can observe load casts that resemble the dendritic flow patterns that one observes flying over Turnagain Arm and Cook Inlet at low tide. While such patterns are common in river deltas, they are somewhat of a surprise at depths of greater than 2000 ft on the ocean bottom. Geologists believe, however, that similar flow mechanisms may explain the resemblance between these two structures. These markings by floral analogy have been called *frondescent structures.* Frondescent casts seem to demand an extremely oozey mud bottom overlying a slope on the ocean floor. As in the above case, a mass of heavier sand sinks into the softer mud; but in this instance, because the mud is in an almost liquid state, the sand continues under the force of gravity to flow down the incline, spreading out into a delta-like, dendritic pattern (Fig-87). Such marks have been observed flowing in a direction opposed to the turbidity current which deposited the sand as evidenced by other sole markings (Dzunski and Walton, 1965). This suggests that the flow occurred after the passage of the turbidity current. Liquefaction of the muddy layer during an earthquake may also trigger the sandy flow long after the passage of the current.

Concretions (Fig. 88)

Sandstone and mudstone beds in Prince William sometimes contain spherical or disk-shaped concretions of limestone. These precipitated from carbonate rich solutions at the time the sandstones were formed. Concretions tend to form around small nuclei of sand grains, bits of organic matter, shells or the skeletal material of marine animals. For this reason, concretions sometimes contain fossil remains. Occasionally, on the southern islands of the Sound, one may find whole concretionary layers of limestone in the turbidites. These often assume a form elongated beds resembling broken statuary. The source of the carbonates forming these layers may be shallow water fossiliferous sediments washed down into the trench by turbidity currents.

Fig-86. Load cast. Note the uneroded mudstone still outlining the sandy knobs. Green I.

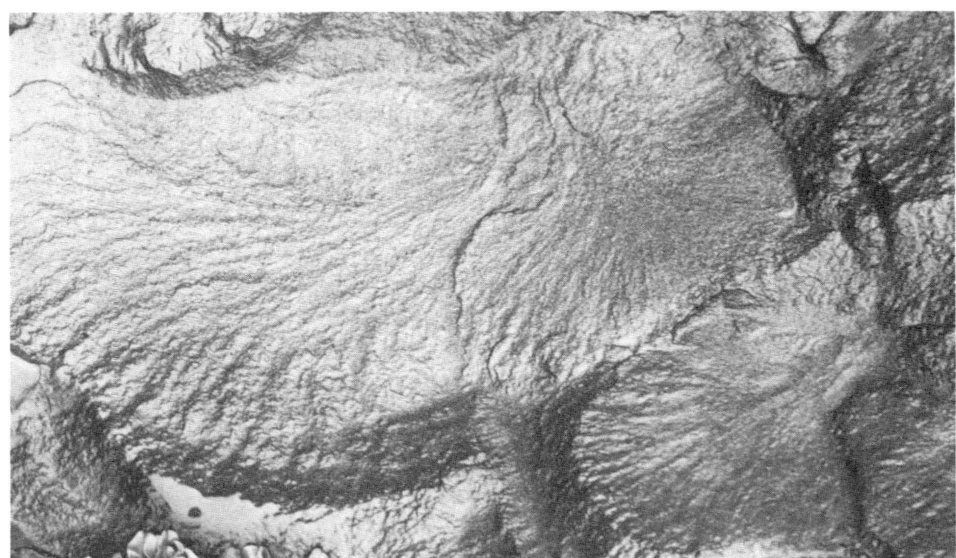

Fig-87. Top. Frondescent marks caused by heavy sand sinking into and flowing through the liquid mud. Note the dendritic pattern. Green Island.   Fig-88. Bottom. Concretion. Limestone concretions form around organic or inorganic material. Montague Island.

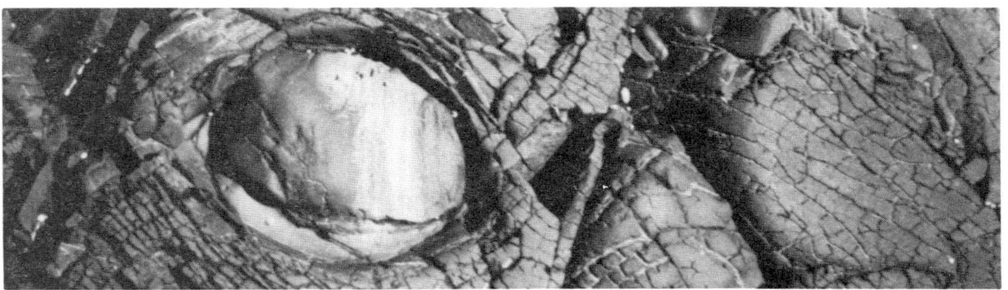

Fig-89. Plant fossils from Green Island.

Fossils (Fig-89)

Sedimentary beds may contain the remains of plants and animals which inhabited the region at the time of deposition of the sediments. These remains, then, provide valuable clues not only to the time of deposition (and hence age of the rocks) but to the prevailing environmental conditions. Unfortunately, the great ocean depths of a subduction trench are a rather sterile environment for most marine organisms. Furthermore, the erosive turbulence of turbidity currents can be very effective in destroying organic remains. And finally, deformation and metamorphism accompanying subduction and uplift can distort or destroy rare fossil remains trapped in trench-fill sediments. Hence, fossils in the Prince William Sound area are very rare indeed.

Fossils, however, have been discovered in the Sound. Tiny, microfossil foraminifera have been found in limestone concretions near Cordova and on

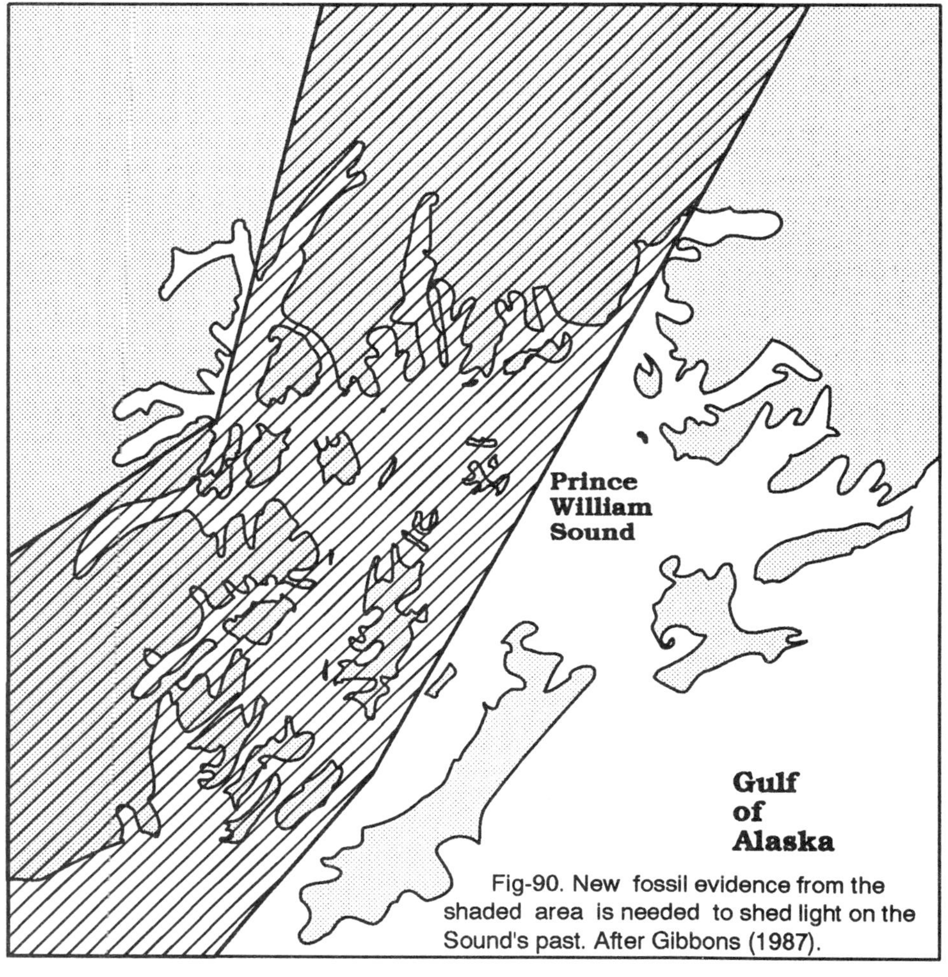

Fig-90. New fossil evidence from the shaded area is needed to shed light on the Sound's past. After Gibbons (1987).

Hinchinbrook Island and radiolaria and other pelagic (deep sea) microfossils have been found on the southwest, outer coast of Montague Island. Fossil crabs, clams and snails of probable Paleocene to Eocene age (Plafker et al. 1985) and worm borings have been found near Galena Bay (Grant and Higgins, 1910), at the Cow Pens Anchorage in Unakwik Inlet and at Bass Harbor on Naked Island (this study). Similar worm borings and several Cretaceous clams have been found in the Port Wells area (Nelson et al. 1985).Two cretaceous ammonites were found at Ziegler Cove in Pigot Bay (this study). A single sea lily fossil is reported from the west side of Evans Island (Johnson, 1918), and poorly preserved conifer fossils were found

Fig-91. Ammonite fossil *(Ammonites Nostoceratidae?)* discovered by author in beach "float" at Ziegler Cove, Pigot Bay. Identification and photograph courtesy of Steve Nelson, USGS-Anchorage.

on the northeast shore of Hawkins Island and at Johnstone Point (Grant and Higgins, 1910). Carbonized plant debris has been discovered at the Cow Pens Anchorage in Unakwik Inlet and on the Northeast end of Green Island (this study). Finally, two different species of Devonian corals (!) have been discovered in cobbles from the Lowe River in Valdez (this study). A marvelous opportunity awaits the amateur geologist in Prince William Sound in the discovery of further fossils. Over the past few years the author discovered several new fossil locations while preparing the field guide section of this book; undoubtedly more remain to be found. Whether the Valdez Group of rocks is really older and different from the Orca Group; whether the Chugach Terrane is really older and different from the Prince William Terrane; whether the Prince WilliamTterrane accreted to the Chugach Terrane during the Eocene or is merely a part of the same accretionary prism; whether the Contact Fault is really a terrane boundary separating rocks of different ages; all might be resolved through the discovery of a single age diagnostic fossil somewhere in the middle of Prince William Sound (Fig. 90).

While doing research for this book during the summer of 1988, the author discovered the fossil ammonite (*Ammonites Nostoceratidae*?) pictured above on the eastern entrance point of Ziegler Cove. This fossil ammonite which is common in the Pacific Northwest confirms that these rocks are of late Cretaceous in age and suggests an affinity with rocks found farther to the south.

Fig-92. Uplifted turbidites on the north side of Passage Canal showing bedding layers.

Bedding (Fig-92).

One of the most prominent features of sedimentary rocks is that they are usually stratified into beds. In the case of Prince William Sound turbidites (or "flysch"), these assume a repetitive and rhythmic character of alternating beds composed of thick and thin strata depending on the massiveness of the turbidity currents that deposited them.

The beds pictured above on the north side of Passage Canal, which are continuous for several miles and which appear to be almost horizontal, are actually somewhat atypical of most sedimentary beds observed in Prince William Sound. Commonly, beds in this region are more discontinuous due to severe folding and faulting. Indeed most are tilted into an almost upright position so that when walking the beaches of the Sound, one often finds himself traversing the up-tilted ends of the bedding. The tilting of these formerly horizontal beds belies their history of being subducted into a trench and gradually, earthquake by earthquake being uplifted into their present positions.

Closer examination of the beds in the above photograph reveal that they actually plunge down into the Earth in a northwesterly direction. Geologists refer to the angle that these strata make with the horizontal (sea level here) as their "dip" and the direction of this plunge as the "direction of dip" (northwesterly here). The trend of these strata along a plane perpendicular to the dip plane extends in a northeasterly direction paralleling Passage Canal. Geologist describe these beds as having a northeasterly "strike" or as "striking northeasterly." Imagine the plunge of the strata as analogous to the down-turned portion of a hinge resting on the edge of a flat surface (such as a table).

The angle that the hinge makes with this surface would represent the angle of dip and its direction would be its direction of dip while the length-wise direction of the hinge parallel to the table top would represent strike.

Often, a close examination of strata in a turbidite bed reveals that the laminae, instead of lying in a perfectly parallel relationship, crosscut one another in a fashion known as cross-bedding.

Cross-bedding is essentially a scour and fill structure; the turbidity current scours out a hollow groove or channel in the existing bottom then fills it in as depicted in the following graphic.

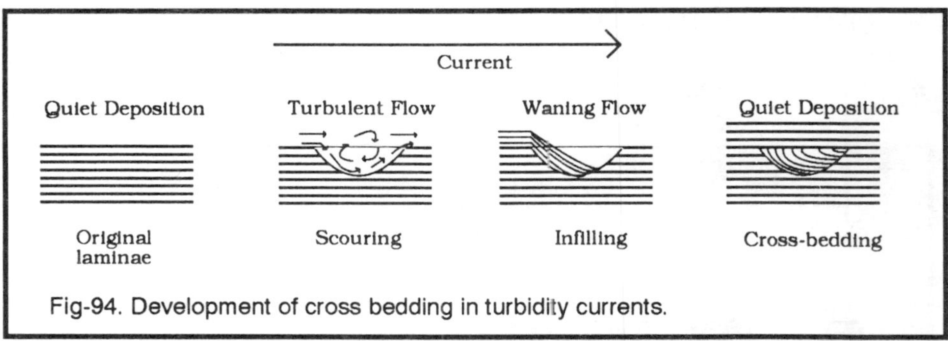

Fig-94. Development of cross bedding in turbidity currents.

Folding and Faulting

Because of their tectonic history of accretion, burial and uplift, most of the sedimentary rocks in Prince William Sound have been folded and faulted. In fact, in some areas the rocks have been so severely deformed that many of the sedimentary features noted above are no longer recognizable.

Brittle rocks tend to react to deforming pressure by shattering and shearing to form faults. More pliable rocks react by compressing, stretching and bending to form folds. Undoubtedly, the heat and pressure of deep burial has a great influence on whether a particular rock reacts by deforming plastically or shattering brittlely.

Folds

Cliffs along the beaches of Prince William Sound often reveal the folded nature of the strata. Geologists classify folds by noting the age sequence of strata in the core of the fold and by noting the relationship of the axial plane of the fold to the planes of its limbs.

The following graphic of a simple fold illustrates some of these concepts. The upward warping strata in the middle is called an "anticline" and the downward warping structures are called "synclines." Since many of the folds geologist examine in the field occur in strata that may be tilted or even overturned, "upward" and "downward" have little meaning; so the geologist classifies the fold as to whether the oldest strata lies in the core (an "anticline") or the youngest strata lies in the core (a "syncline"). Depending upon the competency (resistance to erosion) of the folded strata, synclines and anticlines may determine the shape of the landscape. A syncline with less competent, younger strata in its core may erode away to form a valley while a anticline with competent outer strata may form a hill or the core of a mountain.

The trace of the planes of the fold axes pictured below make the same angles with the planes of the limbs on either side; hence these folds would be classified as symmetrical. Differing angles on either side of an axis would create an "asym-

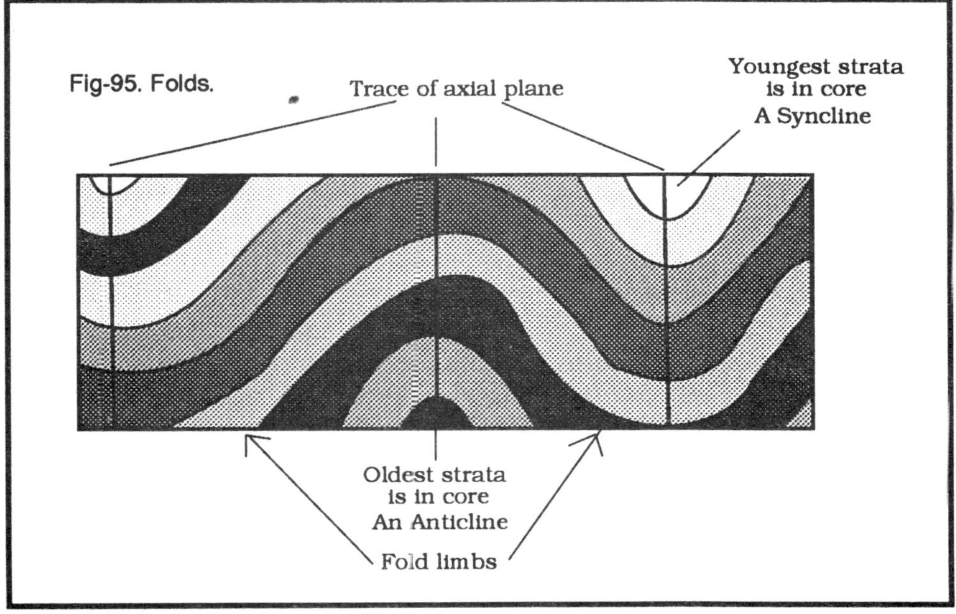

metrical fold." The fold axes here are all parallel; these folds would be classified as "isoclinal." Isoclinal folds are quite common in the strata of Prince William Sound.

In Prince William Sound where the deformational history has been quite complex, few folds are as simple as those depicted above. For example, fold axes are sometimes tilted as pictured below.

Tysdal and Case, mapping the Sound in the late 1970s noticed a curious pattern in the tilt of folds in the Prince William Sound region. Folds to the north and west of the Knight Island ophiolite tend to tilt seaward or to the southeast as would be normal for rocks formed in an accretionary prism. However, south and east of the ophiolite, principally on Montague and Hinchinbrook Islands, the fold axis are overturned landward to the northwest. Geologists, James Helwig and Peter

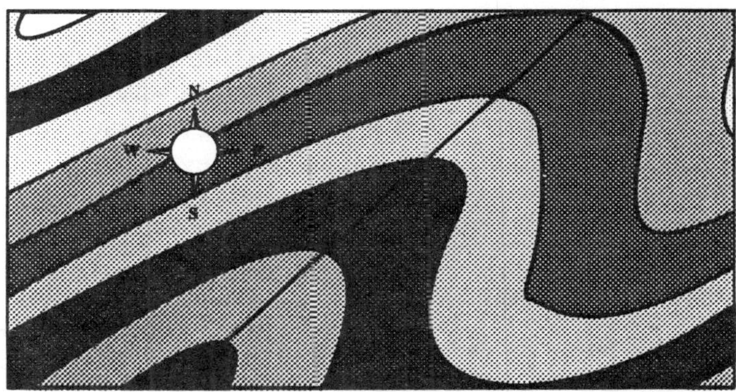

Fig-96. Northeasterly tending vergent folds.

Emmet, in a paper published in 1981 suggested that Knight Island, a chunk of oceanic ridge obducted onto the continental margin, provided a backstop against which these sediments thrust overturning them to the northwest (Cf. Fig-138. Helwig and Emmet, 1981).

Faults

Faults vary from regional features, where entire sections of plates have slipped past or thrust over one another, to small shear zones in local outcrops. Larger faults undoubtedly account for much of the Sound's present landscape. Many valleys in the sound are faults that have provided pathways for stream erosion. Many of the

Sound's deep fiords follow fault zones which provided the pathways for deep-scouring glaciers. On a smaller scale, most of the Sound's mineral wealth tends to occur in or along fault zones.

Most of the Sound's faults are either reverse faults or thrust faults. A thrust fault is actually a low angle reverse fault. Both result from the compressive forces that arise when ocean crust is being thrust below continental crust in a subduction zone. Reverse faults are common in rocks that have undergone accretion in a sedimentary prism. Considerable thrust faulting resulted from the 1964 earthquake – probably because of the low angle of subduction in this area. A reverse fault occurs when the headwall block (the block lying above the fault plane moves up relative to the foot-wall block (the block lying below the fault plane).

Geologists recognize two other general types of faults: normal faults and transverse or strike-slip faults. In a normal fault the direction of slip is in the opposite direction to that of a reverse fault i.e. the footwall moves upward relative to the headwall. Since normal faults tend to occur where the Earth's crust is being stretched rather than compressed, they are rare in Prince William Sound. Strike-slip faulting tends to occur in areas such as Southeastern Alaska and Southern California where crustal blocks are grinding past each other horizontally.

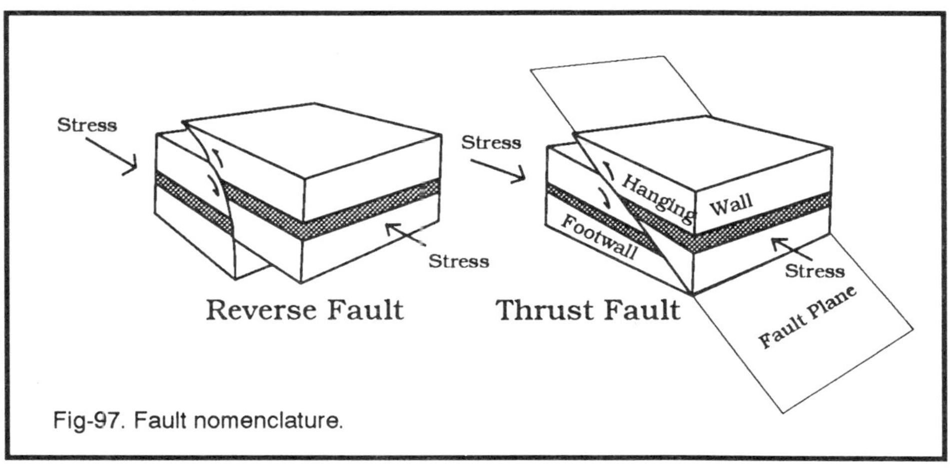

Fig-97. Fault nomenclature.

Clues to faulting on a large scale include offset drainage systems, elongated fiord or valley systems, the juxtaposition of dissimilar rocks, wide areas of shattered rock, and in the case of recent faulting — fault scarps. Fault scarps are easily eroded and the only evidence of a former sharp may be a sudden drop in a drainage system. Such features may also be caused by glacial scouring.

Examining faults as a small scale feature in outcrops can provide some fascinating geological detective work. Close study of a fault may reveal much about its history. First, one should examine the walls of each block; often he will discover groove marks and striations which reveal the sense of direction of movement. A second clue to the direction of movement is to note the direction of offset of strata on each side of the fault. Also, strata near the fault boundary may be bent slightly in a direction opposite to that of the movement of the fault block. Geologists refer to these features as "drag folds." If strata are crosscut by dike swarms or mineralized veins, these may reveal which block has move upward and which downward.

Slickensides, breccia and gouge bear witness to the tremendous frictional forces released during the slipping of a fault. Often, one will find streaking the fault walls a thin, layer of polished rock called "slickensides" which has been formed by the frictional forces of slippage. Between fault walls one may find a zone of powdered and perhaps recrystallized and recemented rock known as "gouge." Similarly, much of the rock in this zone consists of larger, fractured rock fragments ("breccia") shattered during the faulting process. Fault zones often appear as widened cracks in an outcrop because the breccia and gouge are more easily eroded than the surrounding rock.

If a crack in an outcrop reveals none of the above features, then it may have resulted from cooling or other processes and would be referred to as a "joint" rather than a "fault." Both faults and joints may be filled by cooled hydrothermal fluids, consisting mainly of quartz and calcite which sometimes contain gold and other minerals.

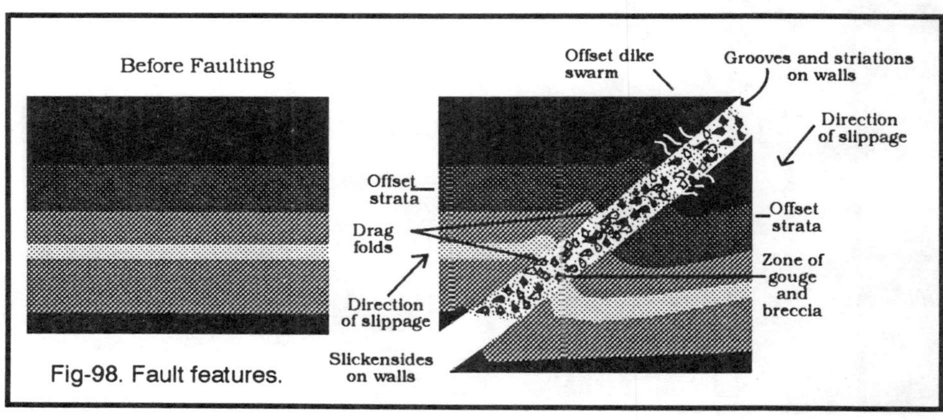

Fig-98. Fault features.

Fig-99. Top. Fault scarp from the 1964 Earthquake. Montague Island. Photo courtesy of S. W. Nelson, USGS. Fig-100. Bottom. Fault fractured rocks along the Landlocked Bay Fault.

Observing Ore Minerals in Prince William Sound

The tailing piles marking Prince William Sound's many mines and prospects offer ample opportunities for the amateur geologists to collect mineral samples. A tool kit for field identification of minerals might consist of a rock hammer (or better yet, a small sledge hammer) for exposing fresh rock surfaces, a hand lens for viewing crystals, a magnet for testing for magnetic properties and a small needle for distinguishing soft flecks of gold from other more brittle sulfide minerals. We will treat only the most common minerals one is likely to discover in tailing piles.

Pyrite (FeS2) is one of the most common ore minerals in Prince William Sound. It occurs in sulfide deposits with copper ores and is commonly associated with gold deposits in quartz veins. The crystals are cube-like, have a metallic luster, and vary in color from a light yellow to brass yellow. These weather to a deeper iridescent or rusty oxide color. As pyrite is often mistaken for gold it is sometimes called "fools gold." However, unlike gold, pyrite crystals are hard and brittle and can just be scratched with a knife or needle. They are non-magnetic. The tailing piles of the Duchess Mine on Latouche Island contains large boulders of massive pyrite. Because the ore in this mine is almost entirely pyrite, the U.S.G.S.investigated the claim in 1955 for its potential for producing sulfur for making sulfuric acid.

Chalcopyrite (CuFeS2) is the most important copper mineral in Prince William Sound. The crystals have a pyramid shape with triangular faces and are slightly darker in color than pyrite and are often of a brassy yellow to a golden color. They have a metallic luster but weathered surfaces may have an iridescent sheen. The crystal are softer than pyrite and are easily scratched with a needle or cut by a knife blade. Chalcopyrite is usually associated with other metallic sulfides in massive deposits and in quartz veins. Chalcopyrite like pyrite is non-magnetic. Samples can be obtained from the tailing piles at Landlocked Bay, and on Latouche and Knight Island.

Cubanite (Chalmersite) (CuFe2S3) is a copper-iron sulfide that is relatively rare most places but fairly common and widely distributed in the sulfide deposits of Prince William Sound. In fact, after chalcopyrite, it is the second most important copper ore in the sound and in the case of the Three Man mine in Landlocked Bay accounted for 23.5% of the copper produced. Cubanite has a slightly more fleshy color than chalcopyrite and is weakly magnetic.

Pyrrhotite (FeS) exhibits bronze colored thin, plate-like hexagonal crystals but

Fig-101. Top. Looking into the adit at Cliff Mine. Mine tailings in front of a mine are good places to look for mineralized rocks. One should not enter adits as many are old and dangerous. In addition, many mines are still privately owned even if they are not being actively developed.   Fig-102. Bottom. Mine tailings at the ore dump of the Standard Copper Mines Company, Landlocked Bay.

usually occurs in massive (non-crystalline) form. Pyrrhotite is slightly magnetic and occurs both in association with the above sulfides in the Sound's copper deposits but also in contact metamorphic veins.

Galena (PbS) is a silver-grey mineral usually found in cubic crystals and a common source of lead. It can only be detected by chemical means in the copper ores of Prince William Sound but appears in visible quantities in gold deposits such as those of the Culross Mine.

Sphalerite (ZNS) like Galena is often present but difficult to detect in the sulfide ores of Prince William Sound. A source of Zinc, sphalerite crystals have a non-metallic, often resinous, lustre and vary in color from yellow to Amber to reddish brown or even black depending on the amount of iron present. Usually sphelerite occurs as zinc-Iron sulfide.

Gold and Silver. (Au, As) Gold is found in hydrothermal veins regionally metamorphosed rocks in the Port Valdez area, in contact metamorphic quartz veins in the Port Wells area, in greenstone at the Culross Mine and mixed with sediment hosted sulfides at other locations. Actually, almost as much gold was produced in the Sound as a by-product of copper mining as all the lode gold mines combined. For example, the Ellamar copper mine produced 51,305 ounces of gold; whereas the Sound's biggest lode-gold producer, the Cliff Mine, produced 51,740 ounces. The Beatson copper mine produced almost one and a half million ounces of silver. Silver in the Sound is often alloyed with gold and has the effect of lightening its color. Most of the gold in the tailing piles of the Sound's lode-gold mines occurs in finely disseminated flecks. The chances of finding a gold nugget are remote. Because of its yellow color and metallic luster, gold is sometimes confused with the above sulfides and sometimes even moist crystals of mica. However, gold's softness and malleability distinguish it. Like Chalcopyrite and mica, gold is soft and easily cut but flattens under pressure; whereas the others are brittle and break up. Gold can be distinguished from pyrite which is hard and brittle. Gold is distinguished from cubanite and pyrrhotite in being non-magnetic. If one places a fleck of suspected mineral on a hard surface and presses on one end with a needle, the hard and brittle pyrite will jump away whereas the softer but brittle chalcopyrite and mica will break into smaller pieces. The gold will ten to flatten or fold around the needle point. Prospector's claim that whereas mica and pyrite lose their luster when moved from sunshine into the shade, gold does not.

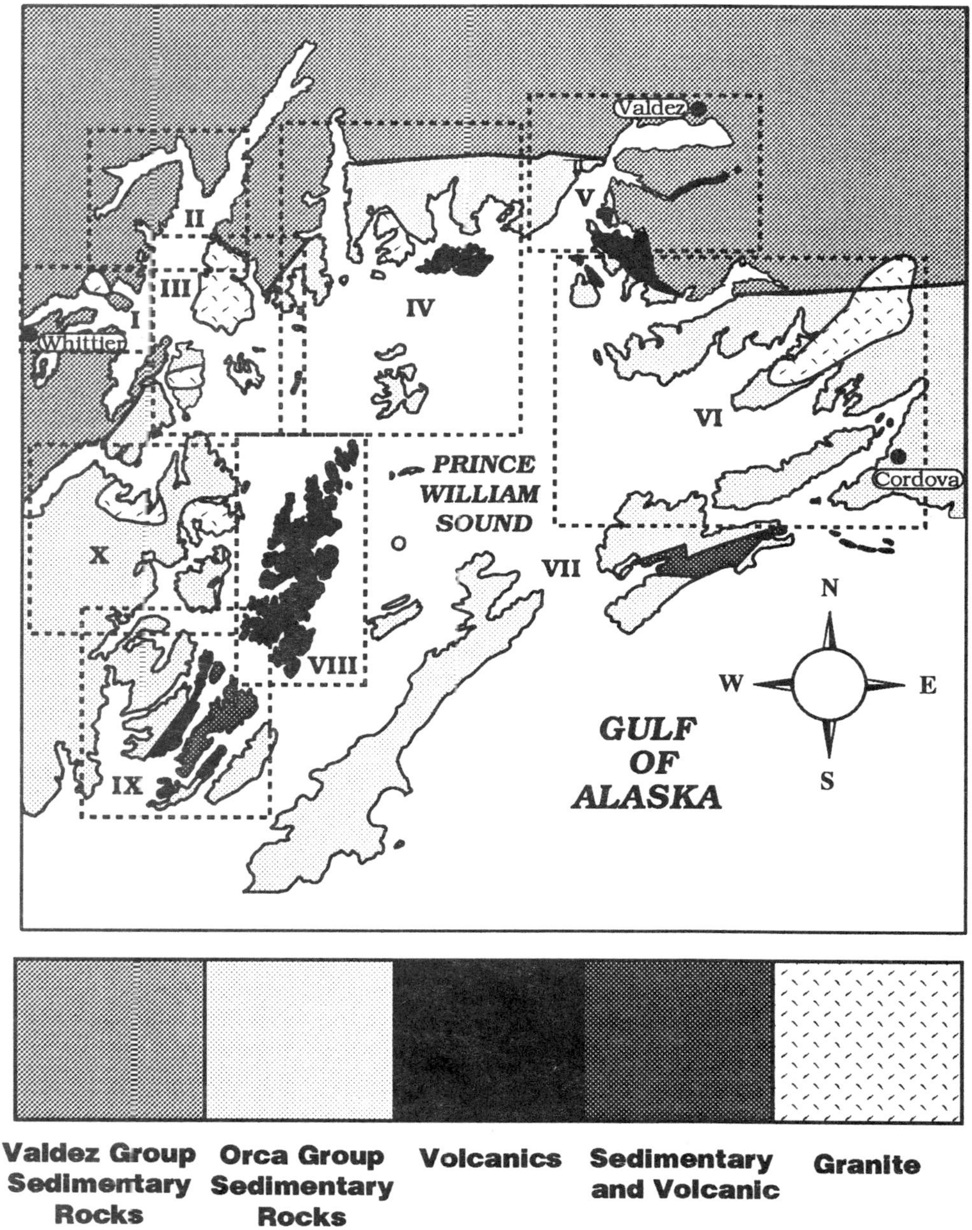

Fig-103. General Geological Map of Prince William Sound. All maps in this section are after Nelson et. al. (1985).

Chapter 4. Field Guide to the Rocks of Prince William Sound

I. Passage Canal Area

General

Except for the small pluton beneath the terminus of Billings Glacier and the larger pluton stretching from Poe Valley to Pigot Bay, the rocks of the Passage Canal Area are mostly weakly metamorphosed Cretaceous turbidites — mostly repetitious alternating beds of slate and graywacke with minor conglomerates in the vicinity of Port Wells. Several quartz diorite or diorite dikes, almost parallel to the bedding and ranging from three to one hundred feet thick, have been noted along the northern shore. Numerous quartz veins cut the rocks at Trinity Point. Aplite dikes, undoubtedly off-shoots of the plutons can be seen on the southern shores. Sedimentary rocks surrounding the plutons have been altered by contact metamorphism. The contact of the Passage Canal Pluton is sharp and the dip of the pluton is shallow so that it crosscuts the steeply dipping country rock.

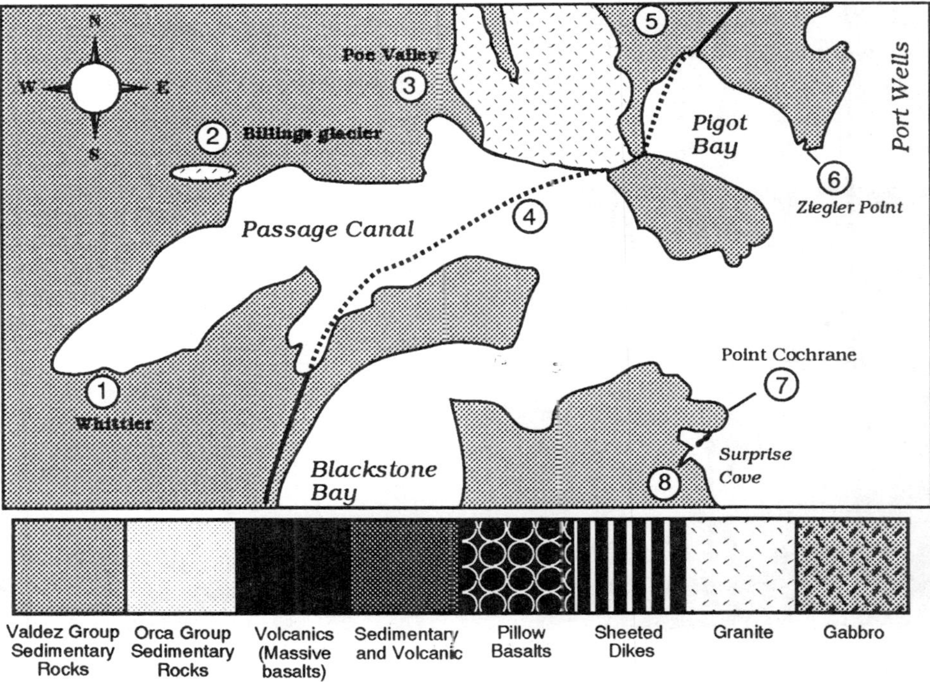

Fig-104. Geological Map of the Passage Canal Area.

1. Whittier is built on the fan-shaped, outwash delta of the creek flowing from Whittier Glacier which perches above the town. Whittier sustained considerable damage during the 1964 earthquake. The area sank 5.3 feet as evidenced by the dead trees at the mouth of Billings Creek across from the town where salt water invaded the roots killing the trees. At least two large breaking waves generated by local underwater landslides devastated the Whittier waterfront killing 13 people. Ten people were lost at a lumber mill which was located on the site of what is now dredged out as the small boat harbor. The first wave was estimated to be 40 feet high and the second 30 feet. Waves may have reached up to 104 feet at other locations in Passage Canal. The Hodge Building sustained moderate damage and was abandoned; whereas, the steel reinforced, fourteen story Buckner Building was only slightly damaged and is still inhabited but has been renamed "Begich Towers." The quake caused an estimated $5,000,000 dollars of damage here.

2. Billings Glacier across from Whittier offers a living geology lesson; for; here, we can view a glacier caught in the act of unroofing a pluton. The light-grey rock visible under the glacier's terminus is granite whereas the darker rocks composing the valley walls are Cretaceous turbidites. The glacier has removed the softer overburden of sedimentary rock exposing the more resistant granite. After viewing Billings Glacier, it is easy to imagine how the granite domes of Nellie Juan or the granite pinnacles of Port Gravina were excavated by glaciers in former times.

The granite here may be a part of the Passage Canal Pluton exposed in Poe Valley a mile and a half to the east. A twelve foot wide pegmatite dike cuts the

Fig-105. Billings Glacier in the act of unroofing a pluton. The lighter rock beneath the glacier is granite; the darker rocks are sedimentary country rock. Photo courtesy of Greg Mallory.

Fig-106. The contact between the country rock and the Passage Canal Pluton is readily visible on the east side of Poe Valley. Photo courtesy of Nancy Simmerman.

granite near the face of the glacier. Quartz crystal up to 18 inches long occur in the pegmatite. Traces of molybdenum, silver, gold, and tungsten as well as some pyrite and chalcopyrite have been found in veins in the pluton and in the contact zone (Jansons et al. 1984).

Above the glacier dipping to the northwest is a thick exposure of classic turbidite sequences. Repetitive and rhythmic beds of alternating light graywackes, dark slates and argillite, some thick and others relatively thin, reveal how massive turbidity flows must have alternated with lesser events as they spilled into the Border Ranges Trench over 70 million years ago at a latitude of 40° north (Fig. 92).

3. Poe Valley marks the western edge of four-mile-long, two-mile-wide exposure of the Passage Canal pluton. This granite intrusion, which has been dated to be about 36 million years old (Oligocene), consists of a core of gray, medium to coarse grained granite grading outward to more mafic and finer grained margins. In places the granite intrudes a gabbro phase (Nelson et al. 1985). Along the granite shoreline of Passage Canal east of Poe Bay, inclusions of sedimentary country rock can be seen in the granite suggesting that the contact zone may lie beneath the bay. The contact zone between sedimentary and plutonic rock is readily visible along the east side and near the head of Poe Valley.

The Portage Bay Mine is located in this contact zone above Seth Glacier at an elevation of 1,150 feet. Passage Canal was called "Portage Bay" when Domenic Veitti and partners located the deposit in 1928. Portage Gold Mines, Ltd. purchased the property in 1933 and between 1935 and 1940 recorded producing 60 ounces of gold and 490 ounces of silver. The gold occurs in conjunction with pyrite, pyrrhotite, and chalcopyrite in quartz in a shear zone. The Bureau of Mines rates the claim as still having moderate mineral potential for a small mine.

4. The Port Wells Fault is a nearly vertical fracture traceable from the southern shore of Bettles Bay to along the eastern shore of Blackstone Bay. The fault is marked by a zone of sheared and broken rock several hundred yards wide. The low col between Logging Camp Bay and Pigot Bay signals the presence of this fault as does the distinctive low notch between the head of Shotgun Cove and Blackstone Bay. At Logging Camp Bay the fault slices though the metamorphosed contact zone dividing the sedimentary rocks of Pigot Point from the Passage Canal Pluton.

5. The Lansing Mine is reachable from the anchorage in the small bight behind the point at the head of Pigot Bay. Follow the stream along the northeast side of the valley until you come to a sawdust pile. Approaching the Passage Canal Pluton, the rocks and cliffs along the route show ever increasing signs of metamorphis. Some ruins above the sawdust pile mark the trailhead which winds up the side valley through muskeg. Alder growing in the muskeg is sometimes the only visible sign of the former corduroy road. A log bridge crossing a narrow gorge signals that the mine is nearby at the bottom of a cliff on the northwest side of the valley. The minerals were most probably in the quartz veins cutting the slate and argillite in this area. Eighty-one ounces of gold and twenty-four ounces of silver were reported from this claim. The Bureau of Mines believes that this mine still has a moderate mineral potential.

Fig-107. Sawdust pile at the trail head to the Lansing Mine.

Fig-108. Felsic dike and chevron folds at the east entrance point to Ziegler Cove.

6. The eastern entrance point to well-protected **Ziegler Cove** exhibits some interesting geological features. The protruding golden-colored rock on the point is a limonite (iron) stained quartz and feldspar dike — undoubtedly an off-shoot from the nearby Passage Canal Pluton. Chilled margins and mineralized hornfels can be found on the east side of the dike. Two similar dikes make off the southeast shore across the bay and another lies farther west along the northeastern shore. This alignment suggests that they may represent two dikes which cut beneath Pigot Bay. To the left of the dike are some chevron folds and at least two small faults cutting the outcrop. A Cretaceous ammonite mega-fossil (*Ammonites Nostoceratidae*) was discovered here by the author in 1988. A second, smaller specimen remains in place in the outcrop at the head of the cove. (Please do not remove). The excellent preservation of both of these as well as several fossil clams discovered on a line between Cochrane Point and Harriman Fiord suggests that this might be a fertile place to search for more fossil evidence which might help to resolve puzzles concerning the Sound's perplexing past.

7. Point Cochrane is an easy walk from the protected anchorage at Surprise Cove. Fossil worm borings have been discovered on argillite bedding planes here, and an *Inoceramus* clam was found on the beach in a piece of argillite float. On the south side of the point is a six inch to two foot wide zone of conglomerate in a fine-grained matrix. A similar outcrop of conglomerate appears on the opposite shore at Pigot Point and at places along the western shore of Port Wells. These probably represent middle fan feeder channels (cf. p. 106 ff.). Near the conglomerate zone just above the beach is a two to three foot exposure of fault breccia embedded in a crumbly iron-stained matrix.

8. Above the **South Arm of Surprise Cove,** a classic shear zone appears to be cleaving the mountain in half. Bureau of Mine scientists explored this shear zone to a point high on the mountain and discovered there quartz-calcite veins containing masses of pyrite, pyrrhotite, sphalerite and chalcopyrite. They classified this unclaimed zone as having moderate mineral potential. Iron stained and slightly mineralized rocks can be found in the rubble at the head of the cove near the bottom of the chute.

II. Port Wells and Harriman Fiord
General

The Port Wells/Harriman Fiord area consists mainly of interbedded turbidites, slates, argillites, graywackes and minor graded conglomerate beds a few feet thick. Rip-up can be seen near the bottom of some of the graywacke beds. Some of the argillite and graywacke beds are thick and massive. Although a poorly defined belt of turbidites exhibiting a slatey cleavage extends from Point Cochrane to Harriman Fiord, the sedimentary rocks of this area have preserved a number of Cretaceous fossils usually found on bedding surfaces of argillite.

Granite outcrops on the western shore of Port Wells, at the head of Pigot Bay (the Port Wells Pluton); and small outcrops occur off both entrance points to Harrison Lagoon's outer bay and on Point Doran. Granite is also visible in the tailings of the Granite Mine and in the tunnel of the Mineral King Mine at Bettles Bay. Many dikes and quartz veins cut the thermally altered rock at the heads of bays along the western shore of Port Wells. Esther Island on the eastern shore is almost entirely granitic. Several four to six-foot-wide quartz diorite dikes cut the country rock on the east shore in the Golden/Avery River area. The mineralization in the area is probably due to the presence of these intrusive masses. Between 1904 and 1916, around 60 mines and prospects were established in the Port Wells/Harriman Fiord area.

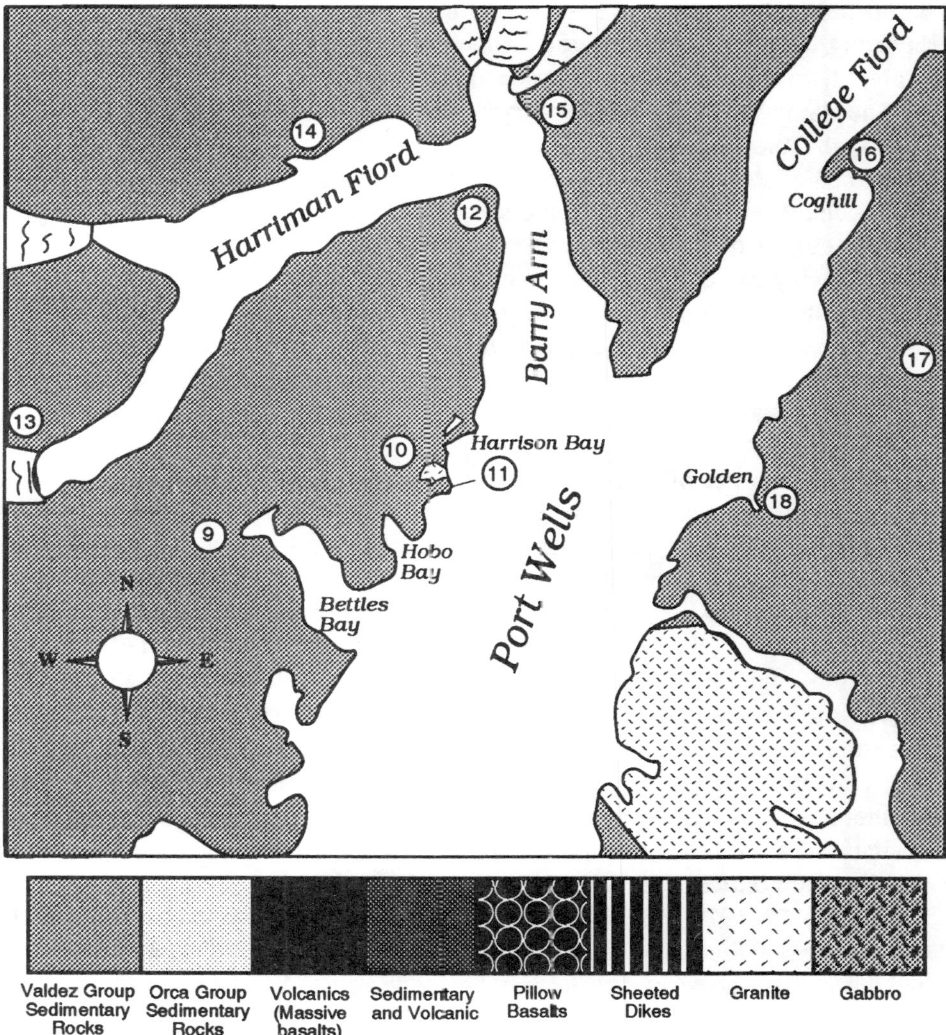

Fig-109. Geological Map of Port Wells, Harriman and College Fiords

9. The **Mineral King Mine** can be reached from the secure anchorage in Bettles Lagoon at the head of Bettles Bay. The stamp mill is visible to the northwest before one enters the lagoon entrance. The stamp mill is best approached first by dinghy at high tide and then on foot on the east side of the stream along the northwest side of the valley. Samples of mineralized quartz can be found in the vicinity of the stamp mill. This stamp mill with a tailing pond below was installed in 1927 replacing the older 10-stamp mill which had been brought over from the Hope mining district around 1920 (Johnson, 1916).

The claim was discovered by George H. Hermann in 1912. In 1920 the Alaska Pittsburgh Gold Mining Co. took possession and began development work. From 1927 to 1939, R. J. Merrill mined the claim intermittently. The years from 1928 to 1933 accounted for most of the mine's reported production of 2,783 ounces of gold and 626 ounces of silver.

Following cables and a wooden hydraulic pipe above the stamp mill, one can climb up the mountain to the actual diggings. The ore deposit occupied a 2 to 6 foot wide fissure containing shattered graywacke and quartz. Much of the gold occurred with metallic sulfides in quartz containing inclusions of country rock (Hoekzema et al. 1987).

10. The **Granite Mine** is best approached from the secure anchorage in Hobo Bay tied to or anchored near the Forest Service mooring buoy. Land on the shale beach just to the right of the buoy. It is an easy walk (about a mile) over the meadows to an overlook of a small lagoon where supplies for the mine were unloaded in former times. When you reach the overlook, head up the meadows toward the woods. You will find a trail which leads up the mountain to the stamp mill. At the stamp mill atop the tailing pile sits an old abandoned truck. Twisted rails lead to a caved-in shaft in the cliff wall behind the truck. If you follow this cliff in a southerly direction for a short distance, you will discover a trail that switch-backs up the mountain to the 350 foot level where there are more tailing piles and another shaft. (**Warning** do not enter old mine shafts; they may be susceptible to caving and thus dangerous to you health.) You can find mineralized quartz, argillite and granite in the tailings here.

On July 19, 1912 M.L. Tatum and Jonathan Erving discovered a mineralized quartz vein 580 feet up on the side of the mountain. Little did they know that they had discovered the second richest gold mine in Prince William Sound. For those of us who have struggled up the mountain just to the 350 ft level, it is hard to grasp how these two old prospectors were able to dig out 10,000 pounds of ore and transport it under winter conditions down to the beach . Yet judging from the heaps of heavy rusting iron machinery lying at remote sites all over the Sound, their efforts were by no means unique for their day.

The workings proved profitable enough that a mill was erected during the winter of 1913-1914. While most of the veins in the Port Wells district began to play out around 1915, the Granite Mine began its period of peak production which lasted until 1922. This is signaled by the moving of the U. S. Post Office from

Golden to this site in 1915 (c.f. Golden below) and the installation of a 10 stamp mill in the same year. The Granite Mine now boasted over 2,000 feet of underground workings and employed a crew of over 50 men between the mine and mill.

The original power for the mine was supplied by a 160-kilowatt, oil burning steam generator on the beach near the small lagoon. A coil of corroding heavy copper wire, numerous insulators, switches and rusting electric motors strewn along the trail to the stamp mill and beyond recall these days. The El Primero Mining and Milling Company took possession of the mine in 1923 and installed a new hydroelectric plant in 1936. Although the mine operated sporadically from 1934 to 1937, 1940 to 1944, and 1963-1964 under a number of different owners, little production of gold was reported (Hoekzema et al. 1987). Most of the ore concentrates were sent to the smelter in Tacoma, Washington.

The gold and silver occur in the presence of metallic sulfides in a 7-inch quartz fissure vein cutting interbedded slates, graywackes, argillite, and granite and a wider three to three and a half foot-wide quartz vein cutting granite and metamorphosed sedimentary rock. In both cases, the quartz contains inclusions of recemented host rock. The granite belies the plutonic origin of the mineralization (Jansons et al. 1984). This medium-grained granite varies from a light gray to a greenish gray in freshly fractured samples but weathered samples have been oxidized to a rich brown color.

The Granite Mine was second only to the Cliff Mine in total gold production in the Prince William Sound area. However, its total production of 24,440 ounces of gold was a little less than half that of the Cliff Mine. This mine produced 2,492 ounces of silver as compared to the Cliff Mine's 8,153 ounces. Presently ,the mine occupies three different levels containing over a mile and a half of underground workings. High grade ore found in the mine and on the property indicates that the mine still has moderate potential for development and the present owner is considering reopening it. The Bureau of Mines rates reworking of the tailing piles as having a high mineral potential. Recreational gold panning in the small streams issuing from beneath the lower tailing pile should reveal some color (Jansons et al. 1984).

11. The beach on Port Wells between Hobo Bay and Harrison Lagoon is a good place to observe outcrops of banded metamorphic rock. The tip of the small point between the Granite Mine lagoon and Harrison Bay is reported to contain a wide shear zone of fractured granite and quartz with collectable 1/2-inch pyrite crystals (Johnson, 1914). The fracture zone contains argillite and graywacke fragments up to 10 ft across and may be a part of the Port Wells fault. Look for outcropping granite in this area.

12. The **Point Doran Peninsula** is reported to contain small irregular outcroppings of granite. Thirteen gold prospects and claims were reported on this peninsula between Point Doran and Bettles Bay in 1914. In-place *Inoceramus* fossil clams are said to have been found on the east side of the point (Johnson, 1914). An 8-36 inch mineralized quartz vein containing lead, copper and antimony is said to be traceable for 200 ft. on the point (Johnson, 1914).

13. A conglomerate bed several hundred feet thick and traceable for over a mile is located above the north side of **Harriman Glacier**. The well-lithified matrix and clasts of this conglomerate are reported to be cut by gold bearing quartz veins (Johnson, 1914).

14. **Serpentine Cove** offers a good anchorage to view some of the typical geology of the Harriman Fiord area. A broad band of black slate can be traced for some distance near the summit of Mount Muir which towers over the anchorage. Two light-colored, fine-grained feldspar dikes can be seen cutting the slate horizontally on both sides of the glacier's terminus at the head of the bay. In the northwest corner of the bay just before the stream, a small feldspar and quartz-filled fault zone contains some good examples of slickensides. Just beyond the eastern entrance to the cove, the remains of the Alaska Homestake mine can be visited. The

Fig-110. Serpentine Cove in Harriman Fiord. Photo courtesy of Nancy Simmerman.

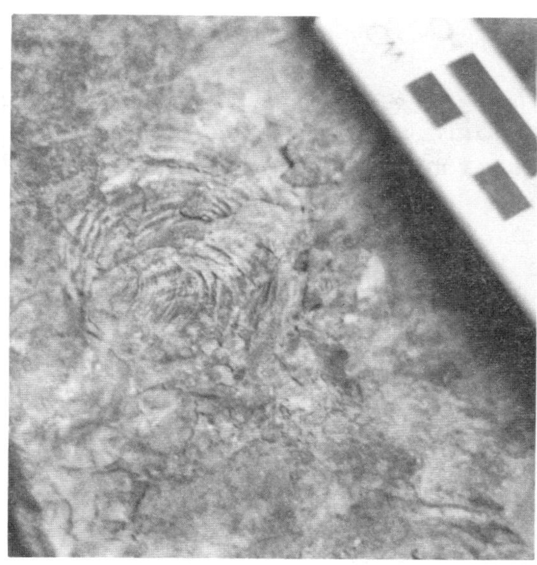

Fig-111. An *Inoceramus* clam of late Cretaceous age used to date the rocks of the Valdez group. Prospectors and geologists have found several in the Port Wells Mining District. Photo by S. W. Nelson, USGS.

Fossil hunters should report their finds to the USGS-Alaskan branch. The USGS allows the finder to keep the fossil after it has been identified.

mine was located in August of 1913. The remains of a 12-ton mill, crusher, and concentrator installed in 1917 lies on the beach. The diggings are in a 6-8 ft. wide felsic dike between 300 and 400 feet on the mountainside above the ruins . A 2-8 inch wide quartz vein in metasandstone along the western wall of the dike contains lead, tin, silver and gold. The mine reported production of 83 ounces of gold and 33 ounces of silver and is still thought to have moderate mineral potential (Jansons et al. 1986).

15. One and one half miles south of **Coxe Glacier** is the site of a former Antimony claim. Grant and Higgins in 1910 reported that 1000 lbs of ore were taken from this claim. A fossil *Inoceramus* clam was also found about 1/2 mile north of this site. Some blue-gray aplite dikes bordering Barry arm are mineralized with pyrite, chalcopyrite and arsenopyrite. Seven prospects and claims occupied the Peninsula between Coxe glacier and Pakenham Point where the mountains seem to be more folded and faulted than those of the surrounding area (Johnson, 1914).

16. Fossilized Worm borings were located by prospectors on **Coghill Point** (Johnson, 1914).

17. The epicenter of the 1964 earthquake has been calculated to lie under this peninsula between College Fiord and Unakwik Inlet. However, there are no surface indications here to indicate the cataclysmic proportions of this event.

Instead, the surface ruptured along faults slicing through Montague Island some 70 miles to the south.

18. It was at **Golden** that a minor gold rush to the Port Wells district occurred. When a one and a half ton quartz boulder containing gold was discovered on the shores of the lagoon in 1911, miners and prospectors from the Turnagain Arm/ Hope gold fields invaded the area from across Portage Pass while hungry prospectors from the Valdez fields arrived by sea. By September of that year, 21 tents and 150 men occupied the spit at Golden. These men fanned out into the hills to establish some 13 prospects and claims in the Golden and Avery River district. By 1913, Golden had become the transportation center and supply hub of the Port Wells Gold District. A 40 ft gasoline launch made weekly trips from Valdez with stops at Wells Bay, Unakwik, Bettles and Hobo Bays. In addition to delivering gold seekers, it carried mining supplies which could be purchased at Golden for prices only slightly above those in Valdez. In addition to mining supplies, the boat carried the mail; for Golden boasted a fourth class post office. However, gold in the Golden/Avery River area proved to be a disappointment compared to what was being discovered across Port Wells on the western shore; and by 1914, the site was all but abandoned.

III. Northwestern Islands

General

This area is comprised of three large islands (Esther, Perry, and Culross) all of which are composed of or include granite intrusions. With the exception of Fool Island, which is granite, and Lone Island, which is pillow basalts, the numerous smaller islands of the group are all sedimentary.

19. Esther Island is a 35 million year old (Oligocene) pluton which exhibits both a granitic and gabbro phase. The gabbro intrudes Cretaceous turbidites and is itself intruded by the granite. The granite also intrudes Cretaceous turbidites. Geologists are unsure of the relationship between these two phases; the granite may represent a differentiate of the gabbro or a melting of sediments by mantle materials. Extending over 50 square miles, Esther Island is one of the larger plutonic exposures in the Sound. Sedimentary rock outcrops on the island at the

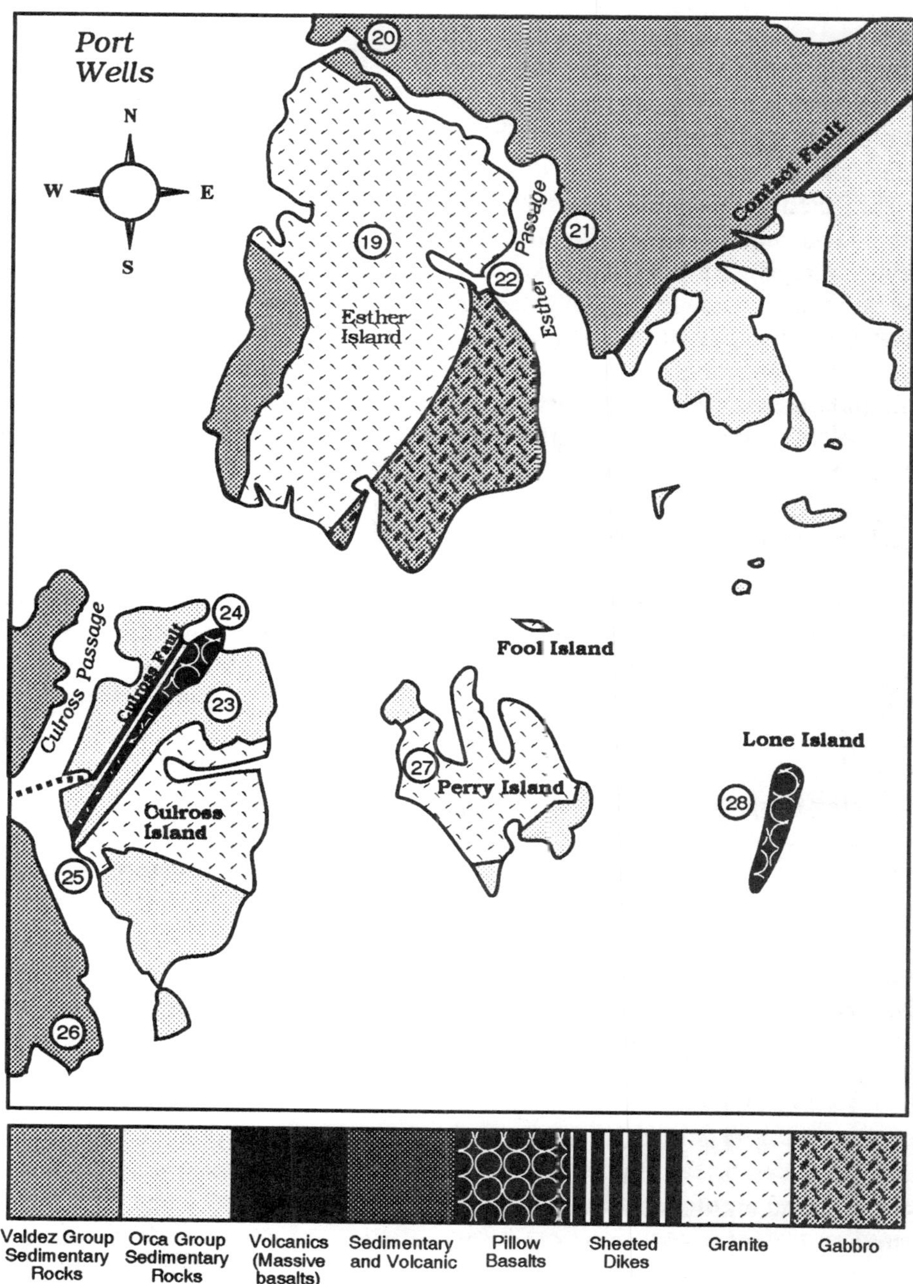

Fig-112. Geological Map of the Northwestern Islands.

northwest end just inside the northern entrance to Esther Passage and along the southwest shore from just south of Granite Bay to Esther Point. Sedimentary rocks on the island and on the mainland shore of Esther Passage exhibit strong signs of both thermal and dynamic metamorphism (Tysdal and Case, 1979; Nelson et al. 1985).

The metamorphic grade of the rocks along the mainland shore of Esther Passage is among the highest in Prince William Sound - especially toward its southern end (Gibbons, 1988). The metamorphic grade increases from north to south along the passage. Esther Passage lies along the contact zone between the Esther Pluton and the sedimentary rock of the mainland and undoubtedly owes its presence to this metamorphosed zone. Anyone who has struck a metamorphic rock with a hammer blow and watched it split cleanly along cleavage planes can easily understand how freeze-thaw action followed by glacial plucking would lead to the excavation of this zone of easily quarried rock.

20. Fisherman's Anchorage is a good, secure anchorage from which to explore the geology of the area. The relatively unmetamorphosed sedimentary rocks of the mainland here display normal turbidite features such as graded bedding, cross bedding, convolute bedding, and scour casts. Most of these features have been completely destroyed by metamorphism farther to the south. However, the beds here are folded and faulted and overturned to seaward (Gibbons, 1988). The Island shore of the anchorage is a good place to observe the contact zone between Esther Island granite and sedimentary rocks. The contact cuts through the valley containing a large lake immediately south of the passage's southern entrance point. Look for offshoot dikes cutting the sedimentary rocks in these areas; some are mineralized containing pyrite, chalcopyrite arsenopyrite and occasionally small garnets. Also, look for inclusions of country rock in the granite near the contact.

21. Banded metamorphic (schistose) rock can be seen along this shoreline. Many of the rocks exhibit a slatey cleavage. Occasional "knotenschiefer"– small ellipsoids flattened in the plane of cleavage which stand out in weathered rock are visible.

22. From the buoy at **Grommet Cove**, one can observe granitic and contact features. On the shore of the thumb at the head of the cove, one can find locally mineralized country rock intruded by granite and cut by quartz veins. Samples of granite and granodiorite, some exhibiting very large crystals, can be found here. One can also examine the contact between granite and gabbro that runs along the small stream with a waterfall at the head of the cove. The steep island cliff dominating the southern entrance to the passage is gabbro and granodiorite.

23. Culross Island exhibits typical turbidite sequences which are locally meta-morphosed, a narrow band of pillow basalts, and a central granitic pluton which is best observed in Hidden Bay. The Culross Mine in Culross Bay is somewhat unique in that it is one of the few gold mines to be found in rocks presently classified as Orca Group and especially as the gold bearing quartz is hosted in greenstone (basalt).

24. On entering **Culross Bay**, one can observe a two-mile long band of greenstone that extends from inside the southern entrance point to within 1/2 mile of the head. The greenstone exists as a distinct ridge of harder rock exposed by glacial erosion paralleling the southeast shore of the bay. This greenstone exhibits pillow forms locally. The pillows occur in tuffaceous beds within the sedimentary sequence. The nearest outcrops of other Orca greenstones occur on Lone Island (13 miles to the east) and on Crafton Island (15 miles to the south).

A major fault, paralleling the greenstone belt, underlies the bay then slices southwestward through Goose and Long Bays and finally under Kings Bay. This fault, which provided a path for glacial erosion, probably accounts for all of these landscape features. Former geologists identified this fault as an extension of the Contact Fault of the mainland to the north (Tysdal and Case, 1979). However, a recent study suggests that the sense of direction of slippage on this fault is quite different from that of the Contact Fault to the north (Gibbons, 1988); hence it is an unlikely candidate for an extension of this propqsed major terrane boundary which supposedly juxtaposes Tertiary Orca and Cretaceous Valdez rocks. Further-more, the metamorphic grade of the sandstones and the sandstone compositions themselves are indistinguishable on either side of the fault (Dumoulin, 1987). No age diagnostic fossils have been discovered in this area to shed light on the respective ages of rocks on either side of this fault.

One can anchor near the head of Culross Bay in front of the ruins of the old mill of the Culross Mine. An old corduroy road leads up the hillside to a lake (150 foot contour) whose stream provided power for the mill. Follow the small streams and ponds feeding this lake to the northeast. You will find steel cables, which formerly supported a tramway to the shore, leading up a gully just above a small beach containing a lot of quartz cobbles. The shaft is at the head of this gully at the 190 foot level in a greenstone cliff. The quartz vein hosting the gold was discovered in 1907 but was probably not mined seriously until 1914 during the mining boom in the Port Wells district. In 1917, a mill was constructed near the beach. Unfortu-nately, the Thomas Culross Mining Company, which had initiated the work and

Fig-113. The eastern shore of Culross Pasage is a good place to view contact features. The lighter rock is aplite, the darker rock is sedimentary country rock.

shipped 12 tons of ore containing some 21 ounces of gold and 19 ounces of silver, went bankrupt in 1918. By 1925, the new owners had produced another 40 more ounces of gold and 34 ounces of silver. Since this time, the claim has been restaked and abandoned several times without any real development work (Kurtak and Jeske, 1986). One can find quartz containing visible gold in the tailing pile in front of the shaft. The Bureau of mines believes that this mine still has moderate mineral potential.

The beach on the northwest shore of the bay opposite the mine exhibits some well preserved sole markings – flute casts, load casts, and ripple marks in slate. Contact metamorphism from the nearby Culross Pluton probably accounts slatey cleavage and baked appearance of these rocks. This beach can easily be reached by hiking over from the head of Culross Cove.

25. The **Culross Pluton** is exposed here. One can anchor near the head of the only cove on the east shore of southern Culross passage. Just north of this cove two waterfalls flow down from a large, classic, amphitheater-shaped cirque carved in the granite. Another wider falls is just south of this cove. The cove lies in the contact between the granite of the Culross pluton and the baked sedimentary rock of Culross Island. The granite near the cove is dark gray weathering to pink or light brown and is fine grained due to rapid cooling at the contact. The granite exposed here is fairly mafic and might be classified as diorite. Where the granite intrudes the slates and graywackes on the cove's northern shore, contact features, especially inclusions of sedimentary rock in the granite, are abundant. Near the head of the cove, many of the shattered boulders reveal a peculiar intermingling of granite and sedimentary rock which represent either granitic dikes in the sedimentary rock or large sedimentary inclusions in the granite. One can see parallel banding and aligned inclusions showing the direction of flow while the pluton was still molten. Near the northern entrance point to the cove, the shoreside cliffs reveal a sharp contact between granite and sedimentary rock. Just south of this location, numerous light colored aplite dikes can be seen cutting the sedimentary rock.

About 3/4 miles north of the cove, a narrow outcrop of light green schist marks the southern end of the greenstone belt extending from Culross Bay to the north.

26. At the **Southern end of Culross Passage,** medium-grained, massive, sandstone is interbedded with finer-grained sandstone and siltstone/argillite. These beds exhibit classic turbidite features such as graded bedding, ripple marks, flute and groove casts, load clasts and rip-up.

27. The **Perry Island Pluton** is best observed from the anchorage in front of the spit in West Twin Bay. This pluton has been dated to be about 34 million years old and is chemically distinct from the other plutons of the Sound. On the north shore of the small cove indenting West Twin's east shore about the middle of the bay, a 40 foot wide shear zone contains stockwork aplite and quartz veins. These have yielded small amounts of tungsten. The contact zone between granite and sedimentary rock is exposed near the end of the northwest entry point to the bay.

28. Pillow structures are clearly visible from a passing boat on **Lone Island** especially near its southwest tip.

Fig-114. The Culross Pluton seen from Hidden Bay, Culross Passage.

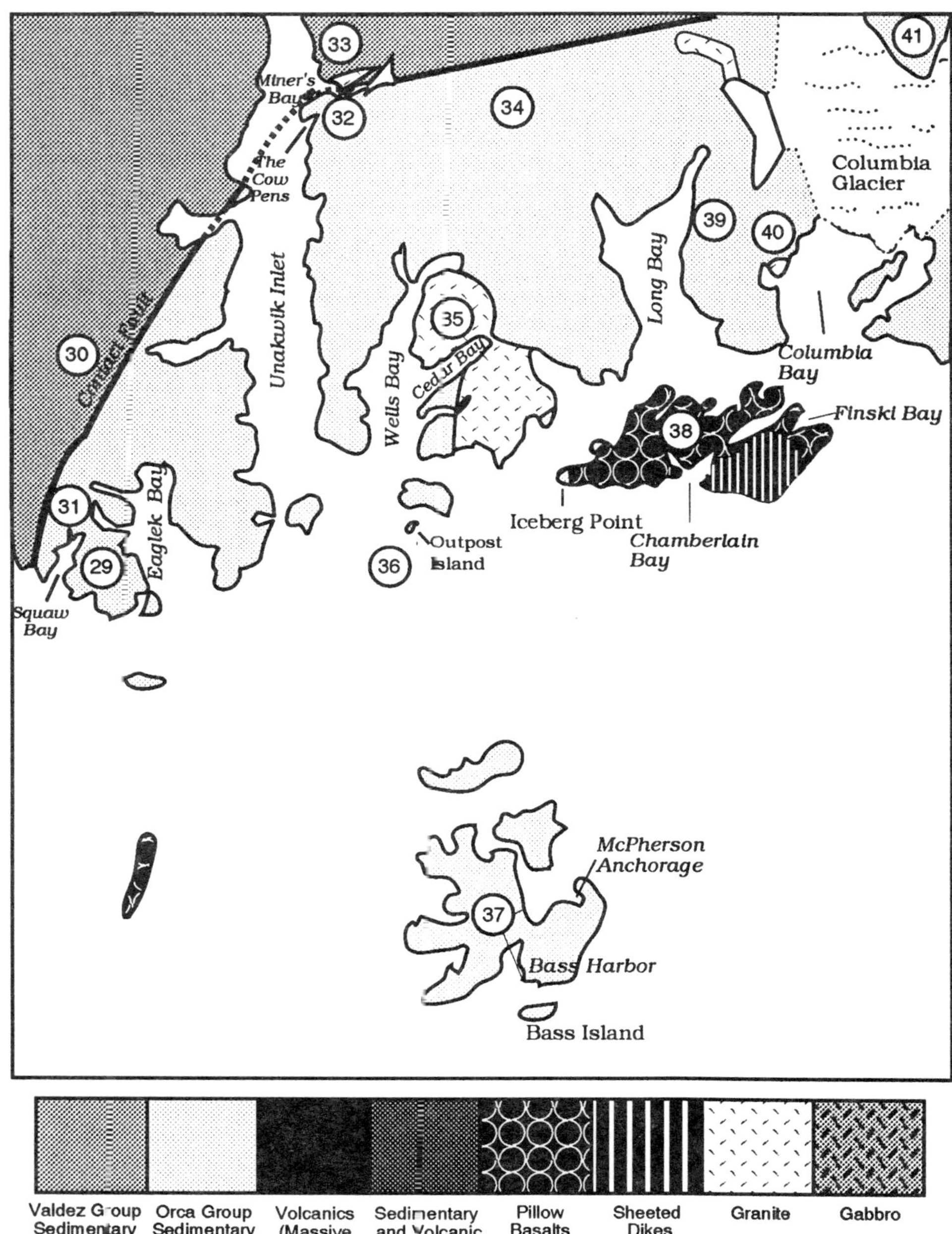

Fig-115. Geological Map of Northern Prince William Sound.

IV. Northern Sound
General

The major geological feature of the northern Sound is a reverse fault which arcs from the vicinity of Squaw Bay northward, skirting Eaglek Bay to cross the mouth of Jonah Bay in Unakwik Inlet, thence eastward through Miner's Bay and then on under Columbia Glacier. This feature has been identified as an extension of the Contact Fault system of Eastern Prince William Sound. Between Squaw Bay and Unakwik Inlet, the Contact Fault places moderately metamorposed turbidites to the northwest over less deformed rocks to the southeast (Gibbons, 1988).

29. The Southeast shore of **Squaw Bay** has good exposures of folded beds; however, these rocks are less metamorphosed than those to the northwest of the Contact Fault. These rocks are a part of a massive sandstone belt that arcs from Icy Fiord to the South to Unakwik to the north. The massive sandstone is medium grained with minor bands of interbedded sandstones and siltstones indicating a probable middle fan origin. Original turbidite features such as graded bedding, cross-bedding, ripple marks, flute and tool marks can be found here and elsewhere in this weakly metamorphosed belt (Gibbons, 1988).

30. The **Contact Fault** is a wide shear zone separating rocks of the weakly metamorphosed sandstone belt from the more highly deformed slate belt to the north. (Tysdal and Case, 1979; Gibbons, 1988). Rocks of the slate belt are metamorphosed sedimentary rocks including sandstones which exhibit a well defined slatey cleavage. In some places banded schists can be found. Mica grains in cleaved siltstone samples often give them a surface sheen. Most turbidite features have been destroyed in these deformed beds (Gibbons, 1988). Although Tysdal and Case discuss the relationship of these two rock groups in terms of a terrane boundary (they, for example, speak of the lower grade metamorphic rocks {Orca} as occuping the "lower plate"), Helen Gibbons interprets the slate belt as sediments underplated at depth at the base of an accretionary wedge. If Gibbons is correct, then Tim Byrne's (1986) model for Kodiak Island may apply here as well. The slate belt (usually identified as Valdez Group) would represent the shortened, underplated sediments, while the less deformed sandstone belt (usually seen as Orca) would represent Byrne's zone of coherent turbidites. The Contact Fault dividing these two groups may be similar to the fault forming along zones of decollement (detachment) between these two groups noted by Byrne on Kodiak (cf. Fig-19). Such an interpretation would suggest an alternative to viewing the Contact Fault in Northern Prince William Sound as a terrane boundary.

Fig-116. A concretion in the rocks along the shoreline at Cow Pens Anchorage. Please leave geological artifacts in place so that others may find and enjoy them.

31. Along the south shore of the anchorage at the head of **Cascade Bay**, one can observe banded metamorphic rock characteristic of the slate belt (Gibbons, 1988).

32. The **Cow Pens Anchorage** just south of Miners Bay is a secure anchorage to explore a number of interesting geological features. The northern shore of this anchorage is composed of conglomerate. The well rounded clasts of this conglomerate are supported in a gritty sandstone matrix. Here, begins the massive conglomerate belt that stretches northward to Miners Bay and along the south shore of Miners Lake. A trip by dinghy northward through the deep channels between a series of long conglomerate islands and the massive conglomerate cliffs of the mainland shore is well worth-while.

A felsic dike about 150 feet wide cuts the conglomerates on the north shore of the anchorage. This dike is probably an offshoot of the Miners Bay Pluton described below. It is composed of rather large crystals suggesting slow cooling. Fault striations are visible in the feldspar where it makes contact with the conglomerates on its west end. Banding on its eastern end exhibits a number of folds.

The steep-to, pinnacle rock in the middle of the anchorage is composed of the same material suggesting that glacial erosion scoured away the softer conglomerates leaving the more resistant dike rock. The long island just south of the anchorage with white feldspar beaches is also composed of the same dike material and suggests a similar origin. The alignment of these three features suggest that they may all represent the same dike. However, similar dike material outcrops at the northern and southern ends of the two conglomerate islands at the western entrance point of the anchorage.

A small stream in the northwest corner of the anchorage divides the conglomerate of the north shore from slate on its eastern shore. Near the stream, spherical cannon ball-like boulders (probably concretions) can be observed on the beach and protruding from the slate which has cleaved neatly around them. Along the same shore, a little farther to the south, one can find sickle-shaped worm casts on the bedding surfaces of the slates. a little farther south near the point, carbonized plant debris can be found in a slatey sandstone bed. This material was most likely carried from shallower water down into a trench environment by turbidity currents.

33. Miner's Bay which lies along the Contact Fault exhibits a number of interesting geological features. Anchorage can be found in the bight just north of the mouth of Miner's River. Just north and west of this bight, one can view the wedge-shaped Miner's Bay Pluton. This pluton intruding slate-belt rocks, seems to be related to the same intrusive event that formed the Oligocene plutons of the northwestern Sound; however, unlike them, its older, approximately 38 million years old, gabbro, mafic phase predominates over the younger, 32 million year old felsic phase which intrudes it. The contact between felsic and mafic phases can be seen on the north side of the pluton. One can find samples of granite, diorite, and gabbro at this site (Nelson et. al. 1986). The Contact Fault forms the southern boundary of the pluton. Both light colored felsic dikes and darker mafic dikes can be seen cutting the sedimentary rock in this area. The remains of a miner's cabin can be found in the northeast corner of the cove and small prospect lies in a shear zone on the north shore just west of the cabin site.

Along the southern shore of Miner's Bay and along the shores of Miner's Lake in the coherent sandstone belt, one can see thick conglomerate beds. One bed is estimated to be 3,000 ft thick (Nelson et al. 1986) and is probably of inner fan origin. This conglomerate belt which stretches on trend with the massive sandstone belt from Miner's Bay to Long Bay exhibits in places chaotic deposits which suggest mass slumping of the inner trench wall (Dumoulin, 1987).

34. A moderately mineralized zone lies just south of the contact zone here.

Fig-117. Top. The Cedar Bay Pluton forms the barren ridgeline above Cedar Bay anchorage.
Fig-118. Bottom. Groove casts in rocks along shoreline opposite Bass Island.

High values for lead, zinc and silver have been found in quartz veins occupying shear zones cutting the graywackes, slate, and conglomerates. Between this zone and the west arm of Long Bay, one vein yielded high fluorine values (Nelson et al. 1984, Jansons et al. 1984).

35. The **Cedar Bay Pluton** has not been accurately dated but is suspected to be of Eocene age because of its chemical similarity to the Sheep Bay Pluton to the south. The Sheep Bay pluton has been dated to be about 52 million years old (Tysdal and Case, 1979). The Cedar Bay granite is of a light gray color weathering to pink, and unlike the Oligocene plutons, exhibits a regular grain size. The contacts with the sandstone and shale are steep and sharp. There are a couple of prospects near the contact zone above the anchorages at the head of Cedar Bay. The first of these is located above the anchorage on the east shore just before reaching the main anchorage at the head of the bay. It is a zinc deposit in a mineralized fault zone in the granite at an elevation 270 ft. and located 1/4 mile northeast of the southern entrance point to the cove. The second is a copper deposit in metamorphosed graywacke on the east shore of the main anchorage at the head of the bay. Both are rated as having low mineral potential.

36. Outpost Island is a tiny outcrop of greenstone suggesting a link between the pillow basalts of Lone Island to the Southwest and those of Glacier Island to the northeast.

37. Naked, Peak and Storey Islands are all sedimentary and consist of alternating beds of sandstone and mudstone with minor conglomerate. Although somewhat metamorphosed and deformed, these coherent turbidites preserve many interesting primary features such as graded bedding, load casts, sole markings and ripple marks. The northern shore of McPherson Anchorage on the eastern end of McPherson Passage exhibits some interesting examples of load casts and rip-up. Quartz clasts in conglomerates on the McPherson Passage side of the narrow isthmus of land dividing the passage from Bass Harbor (west end) are aligned revealing the former direction of the turbidity current that deposited them.

Bass Harbor on the southern end of Naked Island is one of the best places in Prince William Sound to observe turbidite features. One can hike across the narrow isthmus of land from the anchorage in the southwest corner of the bay over to the cliffs and beaches opposite Bass Island or round its eastern entrance point by dinghy. About the middle of of this entrance point, above a prominent outcrop of mudstone, in and above the intertidal zone, is a sandstone face displaying a lattice work of what the appears to be worm borings. Rounding the point, one observes

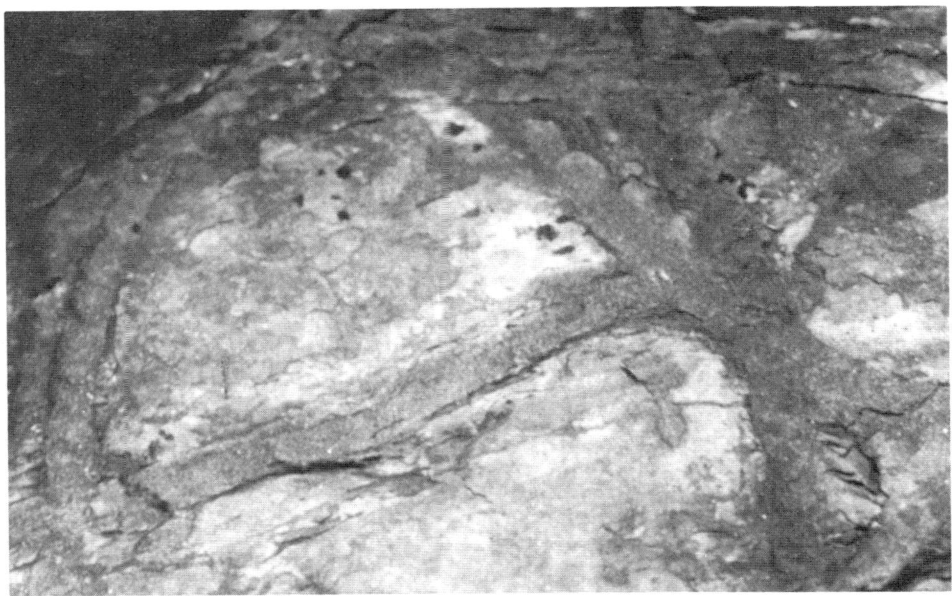

Fig-119. Ancient worm casts left in the rocks along the shore of Bass Harbor.

on the south shore of Naked Island a sequence of upended and folded strata where the mudstones have been mostly or partially eroded away exposing numerous sandstone faces. Examining these faces, one can find some excellent examples of flute casts, groove casts, load casts, rip-up and ripple marks.

On a prominent point jutting from the eastern shore of Bass Harbor near its head, one can find a conglomerate whose clasts have been metamorphosed along with its matrix. The sedimentary, greenstone, quartz clasts in this conglomerate exhibit a definite cleavage while the soft sandstone matrix peels off in thin slices.

38. Glacier Island is composed mostly of pillow basalts and sheeted dikes and is part of a greenstone belt which outcrops first on Elrington and Evans Islands in the southwest corner of the Sound, then trends northward through Knight and Lone Islands before turning eastward near Outpost Island. Crossing Knight Island, these greenstones outcrop again on the shores of Galena Bay and at Copper Mountain. From here, they appear once more near the Midas Mine south of Valdez and at Wortman's Canyon inland east of Valdez. Most of the Copper deposits of Prince Wlilliam Sound are located along this greenstone belt or in the sedimentary rocks bordering it. Geologists have made gravity and magnetic measurements over this belt and have discovered both gravity and magnetic highs following the trend of these mafic rocks. Since basaltic ocean rocks are more dense and contain more magnetic iron than continental rocks, massive quantities of these rocks reveal themselves to these large scale measurements. The high gravity measurements suggest that the basalts of this belt may be as much as 6 miles thick.

It is interesting to note that the gravity high (and trend of mafic rocks that it probably represents) crosses the Contact Fault near the head of Galena Bay placing these rocks in both the Orca and Valdez groups (or in both the Prince William and Chugach terranes, cf. Fig-124). If this belt of rocks is continuous, then the volcanic events they represent must have occurred after the two terranes were joined in the Tertiary. However, if these rocks are continuous and turn out to be older than Tertiary, then it would follow that the Contact Fault cannot be a Tertiary terrane boundary juxtaposing two distinct groups of rocks. Unfortunately, reliable dates for these greenstones have yet to be obtained. However, recent geochemical studies of rare earth elements in greenstones on either side of the contact fault suggest that they may not represent a continuous belt (S. W. Nelson, Oral communication).

Sedimentary rock outcrops on the two most westerly points of Knight Island. On the eastern end of the the most southerly of these (Iceberg Point), there is a showing of conglomerates. Elder and Eagle Bays contain some good examples of brecciated pillow basalts. About 1/8 mile south of Growler Island along the western shore of Growler Bay is a striking exposure of pillow basalts frozen in the act of flowing out onto the ancient sea floor of the Kula Plate. The eastern shore of Chamberlain Bay is a good place to observe sheeted dikes; look for repeated intrusions. Good examples of sheeted dikes can also be observed along the eastern shore of Glacier Island about 1/2 mile south of Finski Point. Eight mineral claims have been staked in and around this sheeted dike region; none, however, have reported any production. The most promising was the Scotia Bell and Portsmouth claims just south of the head of Finski Bay at an elevation of 250 feet. Like many of the claims on Knight Island, which is geologically similar, the small copper deposits are in quartz veins occuping shear zones in the pillow basalts or sheeted dikes.

39. Along the eastern shore of **Long Bay**, one can see chaotic deposits of containing large sandstone blocks embedded in a sandy matrix with little indication of graded bedding. Geologists attribute these jumbled deposits to the mass slumping of the inner wall of the subduction trench (Nelson et al. 1986).

40. A small pluton forms the northern entrance point of **Granite Cove**. Another larger but similar pluton outcrops farther to the north on the west side of the glacier and north shore of Terentiev Lake. These undated plutons are assumed to be associated with the Cedar Bay Pluton which outcrops a few miles to the southwest (Nelson et al. 1986).

41. A small diorite pluton 465 ft. by 56 ft. outcrops on the **Great Nunatak** on Columbia Glacier. Two former gold mines were associated with mineralization from this intrusion – the Gold King and the Ruff 'n Tuff.

The Gold King vein was located on the nunatak in July of 1911 by Olaf Olsen who brought several hundred pounds of highgrade ore into Valdez in the same year. Access to the mine at an elevation of 3,750 was by mule and dog sled from Shoup Bay on Port Valdez; remains of the old telephone line which linked Shoup Bay to the nunatak can still be found on the western shore of the bay. By 1912, a bunkhouse and a mill were built on the nunatak. Between 1914 and 1922, the mine produced 1,997 ounces of silver and 187 ounces of gold. The Ruff and Tuff mine on the south side of the nunatak like the Gold King exploited quartz veins in the metamorphosed slates and graywackes; these quartz veins are undoubtedly off-shoots from the diorite pluton. The Ruff 'n Tuff was supplied by Bob Reeves by air from Valdez. Reeves, who had equipped his plane with skis for landing on Columbia Glacier, would time trips from Valdez to coincide with low tide so that he could land and take-off on the mudflats in front of the town. The Ruff 'n Tuff produced 76 ounces of gold and 20 ounces of silver.

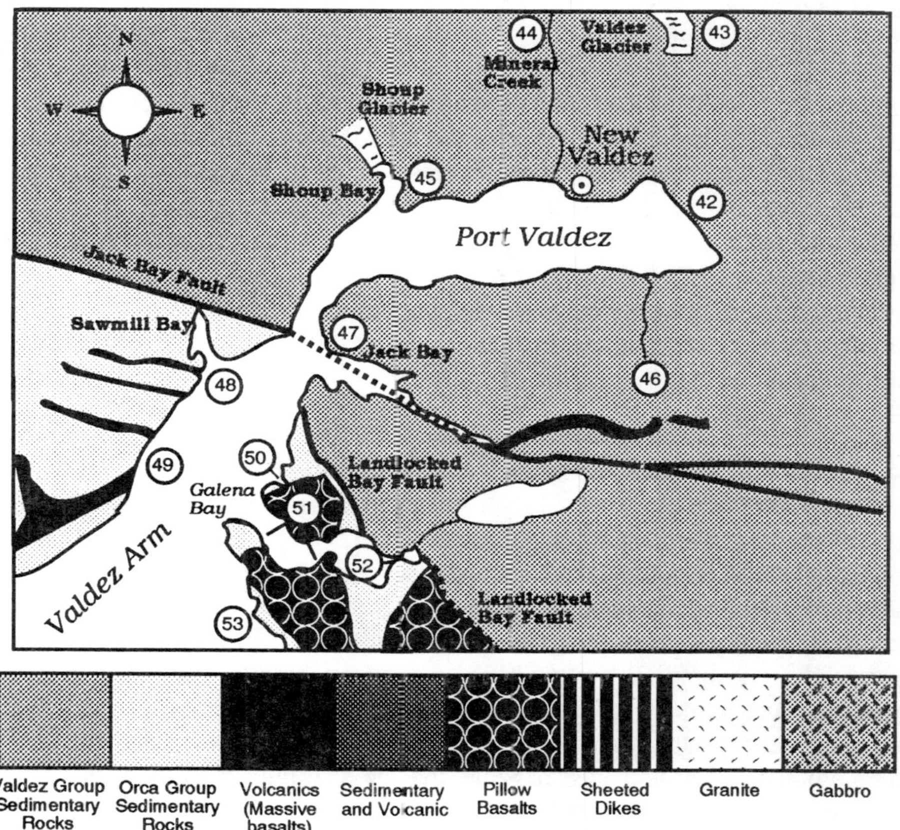

Fig-120. Geological Map of the Valdez Area.

V. Valdez Area

General

Because of its strategic position at the head of Port Valdez near the terminus of Valdez Glacier, a major route to the interior gold fields, the town of Valdez played host to a number of prospectors and geologists in the early days of Prince William Sound's history. When many prospectors of '98 found the arduous route over the glacier not to their liking, they fanned out into the surrounding areas to look for gold. By 1912, forty-eight gold mines stretched over the twenty-six miles between Columbia Glacier and Port Valdez, and over 118 claims had been staked in the immediate Valdez area. In addition, two of the Sound's more lucrative copper mines had been located in the immediate area – the Midas Mine at the head of Solomon Gulch and the Ellamar Mine on Tatitlek Narrows. All of this activity attracted the interest of the federal government which sent U.S.G.S. scientists to map the area and report on the claims. By 1912, over 15 professional papers had been published on the mineral prospects and geology of the area. Among the major contributors were F.C. Schrader (1898), U.S. Grant and D.F. Higgins, Jr. (1909-1910) and A.H. Brooks (1911-1912) . By 1914, nine mills were in operation in the region and 250 to 300 men were employed in mining. Over $900,000 of ore concentrates were shipped to Tacoma for smeltering. Before the end of the mining boom, over 60 claims were staked in the Valdez area alone. Because of its long history of mining and exploration, more is known about the geology of the Valdez area than perhaps any other area of the Sound.

42. Old Valdez was situated on the toe of a loosely consolidated alluvial fan formed by the confluence of the Lowe River and Valdez Glacier Creek where they empty into the eastern end of Port Valdez. The Harbor was built out over a poorly consolidated pile of sediments perched on the edge of a deep submarine U-shaped valley. At 5:36 P.M. on March 27, 1964, the ground beneath the town, which had occupied roughly this location since 1898, suddenly began to shake. Observers described the ground as moving in an undulatory, wavelike motion, rising and falling as much as 3 or 4 feet. The shocks lasted from 3 to 5 minutes. Foundations and concrete structures cracked while wooden structures swayed as if buffeted by some strange, silent and unseen wind. As the shaking progressed, the ground beneath the town shattered into networks of fissures which opened and closed gushing forth small fountains of ground water.

As frightening as these events must have been for those in the center of town, pure and unadorned terror was visited on the waterfront. Here the 400 foot, converted liberty ship, the *Chena,* was quietly unloading cargo before an audience of curious onlookers. As the ground commenced to shake, water began to invade the pores between the loose grains of the sediments beneath the dock, transforming them into a liquid slurry. The entire sediment pile began to ooze forward and over the lip of the submarine canyon. Suddenly, 98 million cubic yards of sediments cascaded over the edge creating a massive turbidity current which would distribute the sediments into a giant fan on the bottom of Port Valdez extending as far the Narrows.

To those on the decks of the *Chena*, these geological events manifested themselves as human tragedy; for they witnessed a large portion of the Valdez waterfront with its living men, women and children suddenly disappear beneath the swirling waters of the Port. Before they could reflect on these awful events, the crew felt the ship lurch sickeningly 30 feet into the air and slam down into the area where the docks and people had been only seconds before. This wave, which was created by water displaced by the sudden slumping, completed the destruction of the harbor and lapped far inland flooding the center of town with 18 inches of water. The *Chena* found herself careened in the mud where the small boat harbor had once been, but the waters rose once again lifting the ship and slewing her around so that she was pointed out into the Port. The Captain ordered the engines to be started and she scraped her way painfully over the muddy bottom and into the safety of deeper water with amazingly little damage. The crew, however was not so fortunate – two men were killed by falling cargo and a third died of a heart attack. Meanwhile, the wave raced to the western end of the Port where after lapping 170 feet up on the mountainside and carrying away the old buildings of the Cliff Mine, it surged out through the Narrows.

Scarcely, 10 minutes after the ground ceased shaking under the town, the backsurge from this wave raged back to the townsite raising a second wave only slightly less menacing than the first. Once again, the center of town experienced some flooding. And this was not the last of the flood waters, for at 11:45 P.M. that night, and again at 1:45 A.M. the next morning, two rapidly moving tidal bores invaded the town as the waters continued sloshing back and forth in the restricted Port.

Many parts of the southern Sound were uplifted during the quake while others in the northwest corner subsided. Valdez was located on the hinge between uplift and subsidence and thus experienced no significant ground level changes.

Devastated by these events but determined to make a new start, the citizens of Valdez moved the townsite four miles to the west to the more stable outwash fan in front of Mineral Creek. Here, a series of bedrock ridges and islands protect it from future surges originating from the head of the bay (Coulter and Migliaccio, 1966). One of these ridges, Dock Point, just east of the Boat Harbor is a good place to see an aplite dike cutting the typical metamorphosed graywacke of the area. At low tide along the seaward side, one can find outcrops of the area's black slate.

West of the harbor at the Civic Center parking lot, are fresh exposures of typical metamorphosed and folded sedimentary beds containing numerous quartz veins. Some of these veins parallel the bedding and are probably older, formed during accretion and metamorphism, while those cutting the strata are probably later and associated with tectonic uplift. It is in these latter veins that most of the gold of the area has been found. Good exposures of quartz veins cutting crenulated beds of slate can be seen in the road cuts between Valdez and Keystone Canyon. The flat, blocky surfaces of the walls of Keystone Canyon display the clean slatey cleavage planes of the area's metamorphosed sedimentary rock. One can see conglomerates just east of the gate to the Alyeska terminal on the south side of the port.

In 1986, Kim Hartman, a local Valdez resident, found two fossiliferous cobbles adhering to the roots of a cottonwood tree he was clearing from his land on the Lowe River Delta. These turned out to be two different species of coral of Devonian age (350 million old). How these fossils which are much older than any others found in the Chugach Terrane arrived in this location is somewhat of a mystery. The closest rocks of this age occur in southeast Alaska in the Alexander Terrane. Because they are composed of a soft carbonate rock, they would probably not have survived glacial and stream transport from older areas of southeastern or central Alaska. Were the Chugach and Alexander teranes once contiguous during part of their strike-slip journey northward or are these native artifacts?

43. The Ramsay-Rutherford Mine, perched 3,500 feet above Valdez Glacier was staked in 1911 and began production in 1914 following the construction of a small mill on the site. In 1915, the mine produced 2,700 ounces of gold and 574 ounces of silver. Intermittent development and production occurred between 1917 and 1925. The mine was reopened in 1934 and produced minor amounts of gold until 1939. The total reported production of the mine during its 25 year history was 5,375 ounces of gold and 1,194 ounces of silver.

Presently, the shafts are all caved in and the remains of the old mill, bunkhouse, blacksmith shop, mess hall and assay office are a pile of rubble on the site. Quartz samples in the area can be found containing disseminated grains of pyrite, pyrrhotite, chalcopyrite, sphalerite, galena and traces of gold and silver (Hoekzema et al. 1987).

44. Mineral Creek just north of Valdez was an important gold mining area between 1910 and the First World War. Several small mills were built in the Mineral Creek area and sixteen claims were active in 1914. The Big Four Mine continued production until 1941, and the Little Giant reported production of small amounts of gold as late as 1955.

One can visit this mining district by following Mineral Creek Road up the Canyon from the northwest corner of Valdez. Driving the gravel road, one first crosses the main creek then continues on crossing three smaller bridges over tributary streams in the next three miles. About 1/2 mile past the third stream, a wide clearing by the side of the road with some rotting ruins marks the location of the old McIntosh roadhouse which serviced the mining district during its height. One mile past the roadhouse, a fourth tributary stream crosses the road and marks the site of the High Grade Prospect. This prospect consists of a series of veins on

Fig-121. The Old Smith Stamp Mill, a two stamp mill, built in 1913 by W. L. Smith to process the ore from his Eldorado Mine. Photo by Glenn Hershberger (1987). Courtesy of Valdez Museum.

both the west and east sides of the creek. The gold is associated with pyrite and galena in quartz. The road ends in another one half mile at an elevation of 550 feet. From here, one can hike the old roadbed along the west side of Mineral Creek for another half mile to Bervier Creek and the old Smith Mill which is now a historical site owned by the state. The two-stamp mill was built in the summer and winter of 1913 by W.L. Smith to process the ore from his Eldorado Mine located at 3000 feet one mile east of Bervier Creek. During the summer of 1914, the mill which employed a minimum of two men processed 120 tons of ore. From here the hardy and fearless bushwhacker can set out for some of the major mines of the district.

The Little Giant group of mines is located about a mile east of the mill at the foot of the glacier in Glacier Creek. Numerous tunnels and diggings can be located in this area. These claims were staked in 1911 and production was reported in 1914, 1917, 1929-34, 1937, 1939, 1948, and 1955. The quartz veins cutting graywacke and slate contain pyrite pyrrhotite, sphalerite, galena, chalcopyrite and free gold. These mines produced 367 ounces of gold and 152 ounces of silver.

Bushwhacking west up Brevier Creek from the mill, one can find the old,

Fig-122. The Cliff Gold Quartz Mine, Valdez, AK. Staked by H.E. Ellis. in 1906. These buildings were destroyed by the 1964 tsunami. Courtesy of the Clifton's Library.

corduroy road marked by alders which leads to the Hercules Mine at the 3,500 foot level. The minerals here are the same as found in the Little Giant group. The mine seems to have been worked between 1912 and 1916. Combined with the Millionaire Mine one mile to the north, the Hercules Mine reported 269 ounces of gold and 440 ounces of silver. Those who would rather not battle the brush can pan for gold just upstream from the mill and where Brevier Creek empties into Glacier Creek (Fechner and Krause, 1987).

45. The Cliff Mine. In 1906, H.E. Ellis of Valdez staked claim to some intensely iron stained quartz veins on the north shore of Port Valdez. If Ellis had known that he had discovered the richest gold deposit in Prince William Sound, it is unlikely that he would have leased the property three years later to the Cliff Mining Company. The Company constructed a three stamp mill on the property in 1909 to be followed by a six stamp mill in 1911. Production began in 1910. The miners discovered that the major ore body angled down under the bay. By 1911 they were digging underneath the bay 100 feet below sea level; by 1913, 8,000 feet of tunnels had been completed — 900 feet of which were 300 feet below sea level. The mine soon became the largest producer in the district. On July 6, 1914 work stopped on the lower levels because of an invasion of sea water. Production continued in the upper levels until 1918. Attempts to pump the lower levels of the mine in 1920 and again in 1933 also failed. The upper levels were reopened in 1936 and mining and milling continued until 1942 when the mining was stopped because of Public Law 208 which closed down mining operations not vital to the war effort. Post-war, fixed gold prices discouraged further development during the late forties and fifties. In 1964, the old mine buildings were swept away by the tsunami. In 1977, a patent for the claim was issued to none other than H.E. Ellis (Hoekzema et al. 1987). During 1987 and 1988, attempts were being made to employ new technology to pump the lower levels of the mine where the highest grade ore had been discovered. As of 1989, the mine had not been reopened. The Cliff mine produced 51,740 ounces of gold – more than twice as much as its nearest competitor the Granite Mine in Port Wells. This amount of gold would have a value of about $19,000,000 at current prices. In addition the mine produced 8,153 ounces of silver.

The gold is found in quartz veins occupying shear zones in the closely folded slates and graywackes of the region. The gold-bearing quartz is usually banded and of a bluish-white color. The gold appears in "fractures, faults, and shear zones along the axes and upper limbs of overturned anticlinal folds." Geologist believe these faults and fractures resulted from dilation "during relaxation of compressive forces of uplift" (Fechner and Krause,198). These fissures were then filled by silica-rich, circulating groundwater heated by the hot metamorphic rocks. These circulating hydrothemal fluids then leached the gold and other minerals from the surrounding sediments. The gold in the quartz veins is associated with pyrite, galena, sphalerite and especially arsenopyrite. By 1914, about 20 mines operated in the Shoup Bay area but none of them produced much gold.

The shoreside cliffs around the Cliff Mine are a good place to observe the intensely folded, fractured, and moderately metamorphosed turbidites of the Valdez group. Numerous quartz veins cut the rocks in this area. On a calm day one can anchor in front of the mine or in windy weather visit the area by dinghy from the nearby anchorage in Shoup Bay.

46. The Midas Mine, located 4 1/2 miles up Solomon Gulch at an elevation of 800 feet, was the fourth largest producer of copper in Prince William Sound. Like the other major producers, the ores of the Midas Mine are hosted in sedimentary rocks spatially associated with nearby greenstones but do not actually occur in the volcanic rocks themselves. Unlike the other major deposits, however; these sulfides are recognized as occurring in Valdez Group rocks; whereas, all the other

Fig-123. Bunkers for the Midas Mine (south side of Port Valdez) owned by the Granby Company of British Columbia. Photo from the Mathilde Gravelle Collection, Valdez, Alaska. Courtesy of Dorothy Clifton.

similar deposits occur in Orca Group rocks. Because the shafts are in good condition and the mines still accessible, geologists have been able to carefully study the ore body in the Midas to gain insights into how this and other sediment- hosted copper deposits of the Sound were formed (cf. discussion below).

The Midas property actually contains two separate deposits — the All-American and Jumbo Lodes. The All-American Lode, located 1/2 mile upstream from the major, Jumbo Lode, was discovered in 1901 by H.E. Ellis who named it "King Solomon's Copper Mines Nos. 1 and 2." In 1904, D.G. Debney relocated the property and moved its boundaries, renaming it the "All-American nos. 1 and 2." Limited development work was done in 1905. The Jumbo Lode, which proved to be the richer deposit, has the distinction of being discovered by a woman; it was located by Mary G. Debney in 1906. After several owners, the property was assigned in July of 1912 to the Midas Copper Co. which shipped about 100 tons of ore to the Tacoma smelter. In October, 1913, the Midas Copper Company sold the property to the Granby Consolidated Mining, Smelting, and Power Co. (Ltd.) of Canada. The new company began actively developing the deposit in earnest. By May, 1914, 1600 feet of tunnels were in place and work on a 5 1/4 mile aerial tram from the waterfront to the mine was begun. One hundred and thirty men were employed on these projects. The tram, whose pilings can still be seen in Solomon Gulch, went into operation in 1915 and was used to carry supplies to the mine and ore to the bunkers on shore. The tram was driven by a large electric motor from hydroelectric power generated at Solomon Gulch as it is today. In 1917, a couple of foot trails and a wagon trail near Fort Liscum connected the waterfront with the mine. Several cottages, a wharf, ore bunkers, a cook and bunkhouse and a blacksmith shop were located on the shore and an air compressor building was located at the mine. A 200 horsepower diesel generator supplied electric power to the mine (Johnson, 1918). The Canadian company sent the ore to their refinery in Anyox, British Columbia for refining Lack of available steamships to transport the ore, rather than a depletion of the ore body itself, led to the closing of the mine in 1920.

In all, the Midas Mine yielded 3,385,680 lbs. of copper (Jansons et al. 1984). Between 1916 and 1919, the mine yielded 2,500 ounces of gold and 15,517 ounces of silver as by products of the smelting of copper (Nelson, 1987). Thus the Midas Mine out-produced most of the lode gold mines of the Sound.

The ore of the Midas Mine consists mainly of pyrite, chalcopyrite, pyrrhotite, sphalerite and galena in massive sulfide lenses or as sulfides interbedded with turbidites. Steve Nelson of the U.S.G.S. suggests that these deposits were probably sulfides produced by black smokers and precipitated out of seawater into small basins on the sea floor at the same time as turbidite deposition (Nelson, 1987). This helps to explain a number of the features of the ore bodies. First, the association of these deposits with greenstones that represent underwater volcanism fits nicely into this framework. The interbedded greenstones and turbidites in the general area suggest intermittent periods of volcanism and turbidity flows. Secondly, the lens shape of the massive sulfide bodies would be expected if the precipitated sulfides filled small rounded hollows in the sea floor and were subsequently buried by turbidity flows. Thirdly, the interbedded sulfides and turbidites suggest that the precipitated sulfides were caught up in the turbidity flows and deposited along with them in segregated, graded beds. Furthermore, geologists studying these interbedded turbidites in the Midas Mine noted that the sulfide beds were folded with the turbidites suggesting that they were laid down at the same time and underwent the same deformations during subduction and uplift. And finally, one can understand the wisdom of an early geologist studying this area when he advised prospectors "The most favorable situations in which to search for copper lodes would seem to be in the black slate and argillite areas, which are in the vicinity of masses of intrusive greenstones (Johnson,1918)." Since slates and argillites are metamorphosed mudstones which would representing the top layer of a turbidity flow, the sulfide precipitates from a black smoker would most likely settle on these layers.

This interpretation of the origin of these sediment-hosted sulfide deposits accords well with the notion that tens of millions of years ago rocks now forming the the Prince William Sound area may have occupied the site of a ridge-trench interaction The spreading center would have supplied the submarine volcanism and black smokers, while the trench environment would have provided the turbidity sediments and accretionary deformation.

Access to the Midas area is from the Dayville Road which leads from Valdez to the Alyeska terminal. Park at the Solomon Gulch Hatchery. Across the road you will find a trail which leads through the woods and up the mountain to the Solomon Gulch Dam. From here the old wagon trail skirts the east side of the lake but is overgrown with alder and some hefty bushwhacking is required. Backpacking a rubber raft up to the lake and rowing to the upper end would be the easiest approach. The lake is now flooded almost to the end of the gulch.

47. The Jack Bay Fault, which cuts across the mouth of Valdez Narrows, divides the rocks of Port Valdez from those of Valdez Arm and is identified as a part of the Contact Fault System. To the north of this fault, the rocks of Port Valdez consist mainly of thick sequences of interbedded slate and graywacke turbidites with minor amounts of arkosic sandstone, argillite, conglomerate and limestone which dip in a northerly direction and have been intensely deformed. The rock is often closely folded with the dominant folds trending in an east-west direction and are overturned to the south. The graywacke is of a bluish-gray color and has been sufficiently metamorphosed in places to exhibit a slatey cleavage and sometimes a crude foliation. The slate is a bluish-gray to black color and has often been metamorphosed to a schist. Both are cut by numerous quartz veins.

To the south and west of the Jack Bay Fault in Valdez Arm, we see the same thick sequences of graywacke and slate turbidites; but they are not as closely folded (Grant and Higgins, 1910; Caps and Johnson, 1915) and generally seem to be less metamorphosed and contain fewer quartz veins than those to the east. Primary turbidite features such as flute and groove casts, absent north and east of the fault, are found in this more coherent turbidite sequence (Winkler, 1976). As the relationship of these two turbidite sequences is similar to that described by Gibbons and Tysdal and Case for the Wells Passage area (cf. the discussion above), it is tempting to see the Port Valdez rocks as the underplated portion of the accretionary wedge and the Valdez Arm rocks as coherent turbidites.

Volcanic rocks, sometimes in pillow form and sometimes interbedded with the turbidites, are much more common in the rocks to the south the fault. Here the arcing gravity and magnetic highs and associated mafic rocks which begin at Elrington Island cross Valdez Arm and continue on into the eastern mainland. On trend with these volcanic rocks, outcrops of pillowed, massive and interbedded greenstones continues *eastward* from the fault at the head of Jack Bay to the area of the Midas Mine in Solomon Gulch above Port Valdez.

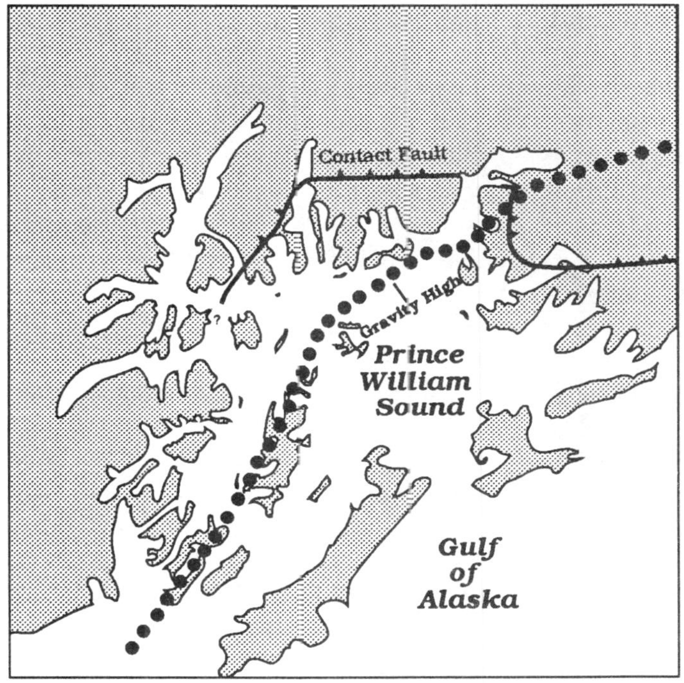

Fig-124. Faults and gravity high. Adapted from Helwig and Emmet (1981).

A third difference between the rocks south and west of the fault is the presence of massive conglomerate beds which can be seen on both sides of Valdez Arm but are most prominent on Rocky Point. These conglomerate beds overlie the greenstone and probably represent inner fan turbidites.

These differences led early geologists to classify rocks north and east of the Contact Fault System as belonging to the "Valdez Group" and rocks to the south and west as "Orca Group." More recent geologists have extended this distinction to argue that the Contact Fault represents a terrane boundary juxtaposing Mesozoic rocks of the Chugach Terrane (Valdez and McHugh) over Cenozoic rocks of the Prince William Terrane (Orca); other geologists disagree maintaining that the two rock groups on a large scale are quite similar and probably represent a single, evolving accretionary prism.

Approaching Jack Bay from Valdez Narrows, one can see mafic dikes on either side of the small cove on the east shore of the Narrows .7 miles northeast of the Entrance Point Light. Rounding the light and heading toward Jack Bay, there is another mafic dike and a small granitic intrusion .4 miles to the south. The anchorage along the north shore 2.5 miles from the mouth of Jack Bay has a good exposure of slate on the north shore. Just east of the point that forms the south shore of this cove is a narrow outcrop of greenstone which extends from the north shore of the bay in an easterly direction for six miles to the vicinity of Solomon Gulch. The Greenstone appears to be intruded into the slate and graywacke and contains inclusions of these rocks, some of which are slightly mineralized. (Johnson, 1918). An extensive exposure of slate forms the east shore of Valdez Arm southward from the entrance of Jacks Bay to Johnston Bay.

48. From the anchorage in **Sawmill Bay**, one can conveniently explore rock formations typical of Valdez Arm. Rocks on the inside of the cove at the southern entrance point are graywacke and conglomerate. The outer side of the point on Valdez Arm is greenstone (look for pillow structures). The inner side of Point Lowe (the northern entrance point) is greenstone overlain by an 800 foot- thick bed of conglomerate. The outer side of Point Lowe is black slate. On the beach on Valdez Arm just north of Point Lowe, one can find samples of graywacke, argillite, black slate and conglomerate. The conglomerate contains angular to subangular clasts of graywacke, greenstone, argillite and slate.

49. On the west shore of Port Valdez opposite the entrance to Galena Bay, there is a good exposure of conglomerates between two outcroppings of greenstone. These conglomerates resemble those of Sawmill Bay to the north and Rocky Point on the opposite shore. These same conglomerates can also be observed on the shores of Emerald Cove in Columbia Bay.

50. Johnston Bay just north of Galena Bay is an important geological site. Four mafic dikes cut the slates at its northern entrance while interbedded greenstones and turbidites make up its southern entrance. Its northern shore is mainly slate and its southern shore mainly greenstone. A mile-long exposure of conglomerate crosses the head of the bay. This conglomerate contains clasts ranging from 8 foot blocks to small granules in a matrix which ranges from a coarse graywacke with calcite cement to a muddy, calcareous matrix. The clasts are angular to subangular and range from graywacke and slate to greenstone. (Grant and Higgins, 1910; Winkler, 1976). Scattered throughout the matrix are calcareous concretions containing nuclei of small fossilized pelecepods, gastropods and crabs. These dark clasts, composed of a dense calcareous argillite, are easily recognized as they are spheroid or oblate spheroid in shape and are more competent than the surrounding mudstone matrix. Their maximum dimensions are about 5 inches in diameter. Approximately 10% of these contain fossil nuclei dated to be of Paleocene or Eocene in age (Plafker and MacNeil, 1966). In 1989, the author found a nodule containing the trace fossil of a clam boring at this site. When geologists discovered these fossils in the 1960's, they revived the notion that the Orca Group rocks might be younger than the Valdez Group rocks and therefore probably constitute a different terrane (the Prince William Terrane). These fossils have been used to establish the maximum age of the Prince William Terrane as Paleocene or possibly Eocene.

On the southern entrance point to the bay just inside the string of offshore islets, one can observe the contact between overlying greenstones and a sequence of well stratified turbidites. A worm cast (*Terebellina Palachei*) was discovered in the interbedded sedimentary rocks (Grant and Higgins, 1910). This trace fossil is not age diagnostic and has been found throughout the Sound in both Orca and Valdez rocks. Massive greenstone flows and pillow basalts can be observed in this immediate area. A little to the south tuffaceous beds and pillow breccias make up the shoreside cliffs – both suggesting rapid cooling of the extruded lava.

51. Galena Bay is the best place to observe geological features characteristic of this part of the Sound. Anchor in the wide bight on the north shore about 1.5

Fig-125. Poorly sorted conglomerates in the Galena Bay bed. Note the "drop hole," upper bedding layers suggesting transport, and lower massive indicating slumping.

miles inside the entrance. From here, one can explore the bay and Johnston Bay to the north. The island which forms the southeast side of the bight has good exposures of pillow basalts. The greenstones of the area display various forms depending upon their cooling histories; some are fine grained due to cooling on contact with sea water; some pillows near the mouth of the bay have been shattered by rapid cooling; elsewhere, the greenstones are massive and have a coarser grain probably due to subsurface cooling.

Overlying the greenstones at Rocky Point across from the anchorage is a 2600 foot-thick bed of poorly sorted conglomerates and pebbly mudstones. The clasts are mostly sandstone, argillite, and greenstone that vary from subangular to rounded in a gritty, graywacke matrix. The clasts vary in size from 8 foot boulders to small pebbles. Some rounded granodiorite and dacite clasts which are foreign to the immediate area can be found. Near tideline, the pebbly sandstones are massive probably indicating a large slump; whereas higher up a crude layering, indicating a certain amount of transport, can be discerned. Geologists believe that these conglomerates represent inner fan deposits that mark entry points of major feeder channels to the submarine fan complex (Winkler, 1976).

Just west of the contact between conglomerates and greenstones (on the west side of the prominent point 0.8 miles from the southern entrance point to Galena

Fig-126. Unusual ripple marks in conglomerate rock at Rocky Point, Galena Bay.

Bay), a large block of conglomerate has separated from the main deposit. The conglomerate surface where the block has separated is ripple marked. The crests of the ripple crests are in places nearly a foot apart and composed of conglomerate with clasts of pebbly sandstone up to two inches in diameter suggesting a tremendous amount of energy in the turbulent flow. Other groove and flute casts can be observed in the turbidites on the south shore beyond the Narrows.

A small cove south of the southern point of the Narrows offers a convenient anchorage to explore the head of the bay. The mountain slopes on the south shore of the cove offer an impressive display of a thick sequence of pillow basalts. Ellamar Mountain above the anchorage consists of several thousand feet of greenstones with interstratified turbidites. A 1680 foot section near Ellamar was found to contain 53% pillow basalts, 39% massive basalts, and 8% intercalcated turbidites (Winkler, 1976). The south shore of the bay offers good exposures of pillows which apparently flowed out onto a soft muddy sea floor interrupting the deposition of sediments. Although the strata have been tilted to nearly vertical, the bottom of the pillows are flattened and lie parallel to the bedding of the underlying slate. The interfaces of the pillows are filled with slate where the soft muds apparently billowed up around the flowing lava before cooling (Caps and Johnson, 1914).

Fig-127. Glory Hole at Ellamar circa 1916. Photo by B. L. Johnson, No. 422. USGS.

The southern tip of the narrows offers some excellent examples of turbidites interlayered with greenstone. Here, the turbidite beds, upended to nearly vertical, are at least 300 feet thick ; the sequence exhibits graded bedding with a bottom bed of conglomerate grading into sandstone which in turn grades into siltstone. These are sandwiched between layers of massive greenstones.

Fig-128. Ellamar Mine, circa 1916. B. L. Johnson photo. USGS.

52. In 1908, the **Galena Bay Mining Co.** constructed a small mining town near the creek at the head of the bay on the south shore. Remains of the old dam, which provided hydroelectric power to the mine 3 miles to the south, can still be seen in the creek. Although an 1800 foot long tunnel was completed in 1909, the mine was abandoned before producing any copper ore.

53. In the summer of 1897, O.M. Gladbaugh and C. Peterson noticed an outcrop of mineralized rock in the intertidal zone in a small bay on the north shore of Tatitlek Narrows. This outcrop was later to become the **Ellamar Mine**, the second largest producer of copper in Prince William Sound. The first ore was shipped in 1900 when 225 tons were blasted from this outcrop. A shaft was constructed inshore of the deposit in 1901. In 1909-1910, a cofferdam was built around the ore body in the intertidal zone to exclude seawater; a 200 foot-deep glory hole was dug behind the protection of the cofferdam; shafts tunneled under the bay from the bottom of the glory hole. By 1912, a large wharf extended from the glory hole and out into the bay while a number of buildings lined the shore. Pilings of this old wharf can still be seen in Virgin Bay. In 1916, 100 men were employed for the 355 days of operation of the mine and a small town stood on the site. The mine reached a depth of 600 feet with 9,300 feet of workings on eight levels. The ore was finally exhausted in 1919; the diggings were allowed to flood; and the buildings were converted to a cannery (Moffit and Fellows, 1945-46). In all, the mine produced 15,761,337 pounds of copper, 51,305 ounces of gold, and 191,615 ounces of silver. Thus, the Ellamar mine produced as a by-product of the smelting of copper ore only slightly less gold than did the nearby Cliff Mine.

Like the Midas Mine and the other big producers of copper in Prince William Sound, the ore of Ellamar Mine is hosted in turbidites associated with greenstones. The main lode was a large lenticular body of massive sulfides with a greatest diameter of 240 feet and a least diameter of 90 feet separated by a thin bed of slate from an overlying lenticular mass of pyrite 35 feet thick. The major copper ore is chalcopyrite with significant quantities of cubanite and sphalerite. The ore body is hosted in slates cut by greenstone dikes and is associated with a dark magnesium-rich limestone – probably a byproduct of the interaction of basic lava with the minerals in sea water.

General Scenario

The rocks of Valdez Arm tell a story of early Tertiary deposition of gravels, sands, and muds by turbidity currents in a subduction trench – probably near a former island arc terrane dissected to its plutonic core (perhaps the Wrangellia-Peninsular Superterrane). Later, during the Eocene, pillow basalts began pouring forth through a rift in the sea floor. Periods of submarine volcanism alternated with turbidity flows creating the intercalcated deposits of greenstones and turbidites that make up Ellamar and Copper Mountains. The intermittent eruption of basalts may have been quiescent for extended periods for the interbedded sediments attained thicknesses of 300 feet or more in certain places (Moffit, 1951).

The interaction of the Kula Ridge with the Border Ranges Fault may account for this episode. Like its modern analog, the Juan de Fuca Ridge, black smokers along the rift seem to have been spewing forth black clouds of sulfide-rich smoke which settled into hollow depressions in the muddy sea floor forming the copper deposits of the Midas and Ellamar mines. These volcanic deposits were later covered by inner fan conglomerates and slump deposits as they approached the inner trench wall on the conveyer belt of the Kula Plate.

The sediments which formed the rocks north and east of the Contact Fault system (Valdez Group) were probably those underplated during subduction when they were shortened, folded, faulted and metamorphosed. The turbidites, greenstones and slump deposits south and west of the Contact Fault System (Orca Group) were most likely part of the coherent middle portion of the accretionary wedge which was less deformed during subduction.

From the Miocene through the Pliocene and probably into the present, these buried trench-deposits were uplifted, further deformed and metamorphosed. During the Pleistocene, glaciers sculpted and carved the uplifted strata into their present forms. Signs of glaciation high on Ellamar and Copper mountains suggest that the ice at one time reached the 3000 foot level. With the melting of the glaciers in the Holocene, U-shaped valleys were flooded forming today's intricate fiord system.

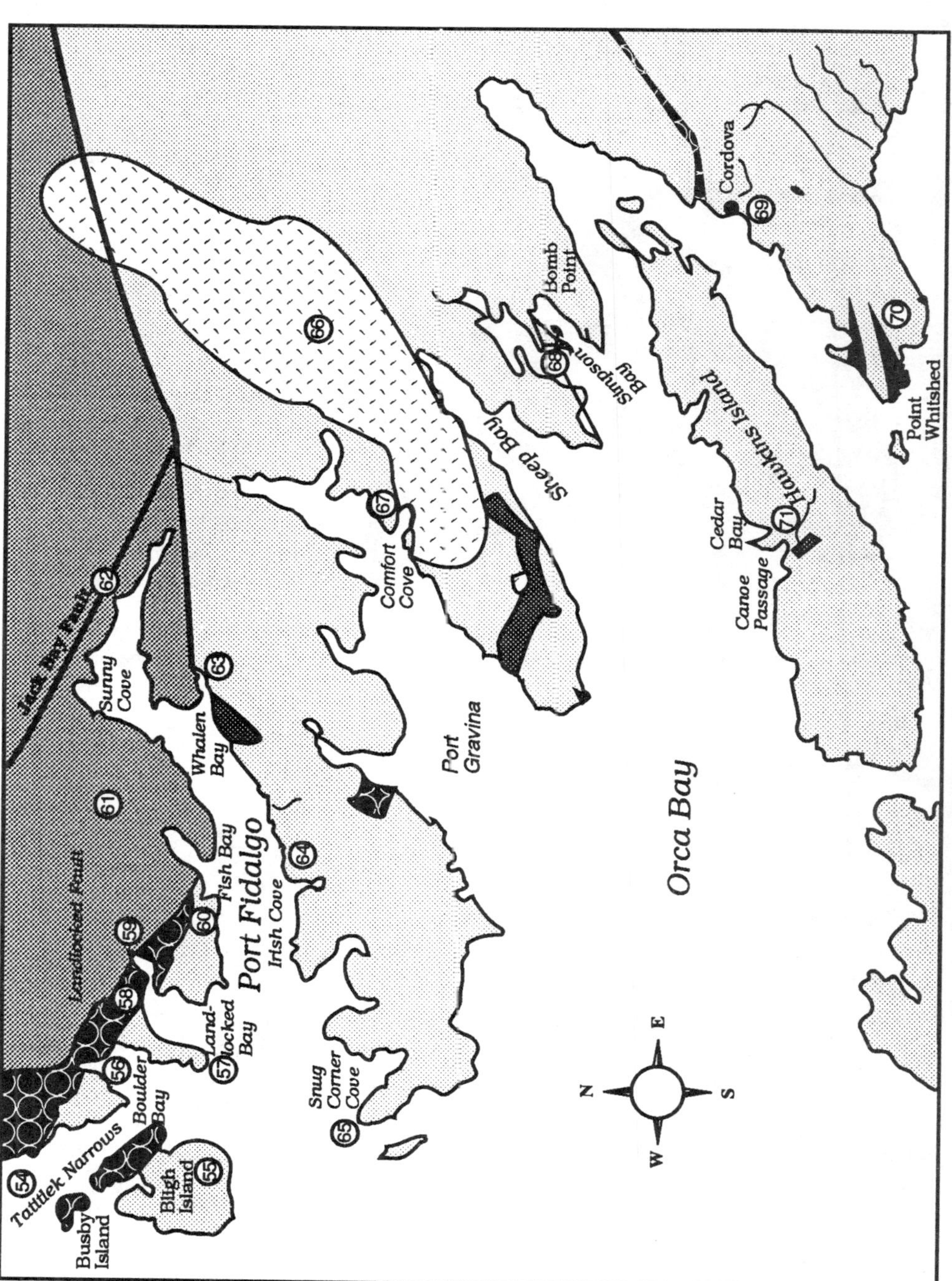

Fig-129. Geological Map of the Cordova Area.

Fig-130. Copper Mountain is composed of interlayered turbidites and greenstone hosting copper deposits. Tatitlek Village lies in the foreground. Photo courtesy of Nancy Simmerman.

VI. The Cordova Area

General

The Cordova Area, as defined here, is bounded on the north by the Contact Fault system consisting of the Jack Bay, Landlocked and Gravina Bay thrust faults. The rocks of this area consist mainly of deformed Orca Group turbidites. In places, massive greenstones and pillow basalt flows are interbeded with the turbidites. While significant deposits of metallic sulfides have been discovered in the Port Fidalgo area where these two rock types are in contact, similar contacts on Hawkins and Hinchinbrook Islands and in the Port Gravina and Cordova areas seem to be barren. The Eocene Sheep Bay Pluton intrudes this area. Scattered microfossils collected from scarce limestone concretions in this area suggest Paleocene/Eocene dates.

54. Tatitlek Narrows appears to have been formed by glaciers bulldozing out a bed of softer slate overlying more resistant greenstones. Rocks on the mainland side of the passage near the village of Tatitlek are alternating beds of graywacke, thick black slates, and thin beds of conglomerate which have been extensively folded and faulted. The slates, here, have a well developed cleavage which is difficult to distinguish from bedding.

55. The southwestern (**Bligh Island**) shore of the narrows is composed of interlayered pillow basalts and sedimentary rocks as is Busby Island to the north. Bligh Island southwest of a line from the head of Cloudman Bay to the head of West Bay is composed of turbidite sequences which exhibit more open folds than most of the strata of Prince William Sound. Primary turbidite features such as sole markings can be observed in the strata forming the western entrance point to West Bay. Gold has been found in quartz veins in greenstone near the entrance to Cloudman Bay and on the shore facing the narrows opposite the island near the north end. There is an old copper prospect at the head of Cloudman Bay.

56. In 1910, the Reynolds-Alaska Development Co. held 20 copper claims at the head of **Boulder Bay** where there were over 2100 feet of workings drilled into Copper Mountain. A wharf, electric plant, air compressor, office, bunkhouse, warehouse and superintendent's house stood on this site.

57. Bidarka Point, between Boulder and Landlocked Bays, is composed of a thick, three mile-long bed of black slate. The slate has a well defined cleavage which may be mistaken for bedding. Glacial till is observable on the southeast side of the point. Just beyond where the slate makes contact with the greenstones on the west shore of Landlocked Bay, one can find basalt pillows whose forms are outlined with green chert.

58. Landlocked Bay in Port Fidalgo is one of the best places to collect samples of copper ore and examine relics of the Sound's copper mining era. It was here in 1897 that the first copper claim was staked in Prince William Sound by the Alaska Commercial Company which ran the trading post at Port Etches on Hinchinbrook Island. Latouche and Ellamar were staked later in the same year. Serious production of and prospecting for copper began around 1900 and peaked in 1907. Most of the activity was centered in the Ellamar, Port Fidalgo, Knight Island, and Latouche Island areas. Early shipments of high grade ore to the smelter in Tacoma, Washington averaged as high as 10% copper. Production continued throughout the teens at the big-three mines at Solomon Gulch, Ellamar and Latouche Island and sporadically at other locations. The financing of the First

Fig-131. Top. The Threeman Mine at the head of Landlocked Bay circa 1912. Photo by B. L. Johnson (440), USGS Photographic Library.
Fig-132. Bottom left.Tailing pile from the Threeman Mine. (1989).
Fig-133. Bottom right. Collapsed adit at the Threeman Mine. (1986).

World War attracted capital away from speculation on copper mines spelling the death knell for further prospecting in the Sound. Between 1900 and 1930, 214 million pounds of copper were produced by fifteen companies. Approximately 96% of this production came from Latouche and Ellamar. The closing of the Kennecott properties on Latouche in 1930 signaled the end of copper mining era. Since that time, several surveys by private companies, the U.S.G.S. and the Bureau of Mines have attempted to establish reserves for possible future exploitation. About 23.5 % of the copper from Landlocked Bay occurred in the mineral cubanite (chalmersite).

58a. The **Threeman Mine** is located on the north shore opposite the anchorage at the head of the inner bay. Reddish colored tailing piles on the shore mark the location. One can climb the tailing piles to two adits on either side of the tailings at about the 200 foot level. Excellent samples of chalcopyrite, pyrite and cubanite can be collected from the tailings. The Threeman, located in the Landlocked Bay overthrust fault zone was the major producer in Landlocked Bay. Staked in 1903, the mine produced a small shipment of ore in 1904 and made regular shipments of high grade ore from 1912 to 1915 when the mine was closed. Between 5,000 and 6,000 tons of 8% ore were shipped from this mine which includes over 5,000 ft of workings on 5 different levels. Most of the ore was taken from two massive sulfide lenses hosted in greenstone, slate and graywacke. Like the Ellamar and Beatson claims, the ore is located in a shear zone. This led early geologists to speculate that the sulfide lenses were deposited by hydrothermal processes. However, if the modern interpretation that these deposits are the result of submarine hot spring activity subsequently buried by turbidity flows is correct, then it is more likely that later faulting occurred along the sedimentary/greenstone contact and may have been guided by the lack of competency of the massive sulfide bodies. Slickensides in the Ellamar mine suggests such an interpretation.

58b. The **Landlocked Bay Mining Company** established a claim on the south shore of the inner bay in 1898. The old dock is still visible near the middle of the cove. Just east of the dock is a modern cabin near some reddish tailings on the beach. If one rummages around in the woods behind the cabin, he can find the old rails disappearing into the caved adit. Major work on the claim was performed in 1906 and a small quantity of 7 1/2% copper ore was mined here.

58c. Across the bay from this mine, some diggings in a mineralized shear zone are visible. Look for malachite staining high on the cliff side. This is the **Hemple Prospect** which consists of four adits dug in 1915.

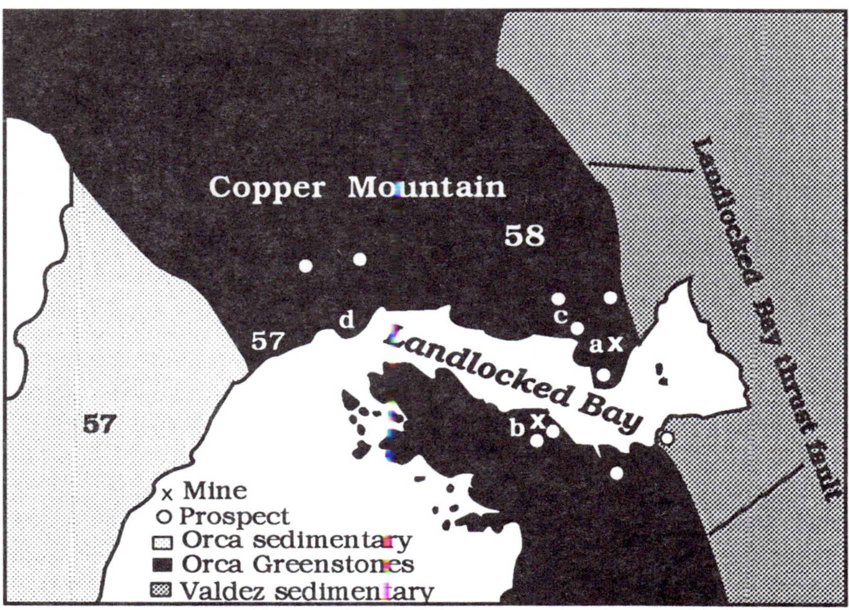

Fig-134. The Mines and prospects of Landlocked Bay

58d. The **Standard Copper Mines Company** ore dump is visible on the point at the narrow entrance to the inner cove. This ore was brought down to tideline by a 2,526 foot aerial tram connected to the mines at the 2000 foot level on Copper Mountain where there were 5 tunnels and 1300 feet of workings. There were several buildings on the beach near the dump and at the head of the tramway high on the mountain. Several tons of high grade ore were removed from sulfide lenses in sheared greenstone and shipped to the Tacoma smelter between 1906 and 1911.

59. The **Landlocked Bay Fault** crosses the head of Landlocked Bay. This fault overthrusts Valdez group metasedimentary rocks above Orca greenstones. Look for a several hundred foot wide zone of shattered rock near the sedimentary/greenstone contact at the head of the bay (Fig-100). Some geologists believe the fault represents a portion of the Contact Fault System and hence a terrane boundary between the Cretaceous Chugach and the Tertiary Prince William terranes.

60. Conglomerates can be viewed on the western entrance point of **Fish Bay** (Winkler and Tysdal, 1977).

61. For those willing to make the arduous five-mile trek from the anchorage at **Sunny Cove**, samples of ultramafic rock (soapstone) may be obtained. In 1983, U.S.G.S. scientists discovered a rounded hill of serpentized dunite and peridotite 700 foot in diameter. This ultramafic bodies located at the 2500 foot level at the toe of the glacier above the valley northeast of the anchorage. These mantle rocks formed deep within the earth seem to have been squeezed up into the Valdez turbidites (Nelson et al. 1985).

62. A lenticular lens of marble (metamorphosed limestone) some 12 feet thick overlies the surface of pillow breccias at this location (Winkler and Plafker, 1981).

63. The **Fidalgo Mining Company Mine** consists of four adits in interbedded slate, graywacke and greenstone on the mountainside just south of the entrance to Whalen Bay. Massive chalcopyrite, pyrrhotitie and pyrite occur in 1 to 5-foot-wide veins. A total of 360,376 pounds of copper and 12 ounces of silver were taken from these workings.

64. Two copper mines are located on the mountainside just east of **Irish Cove**. Fallen down mining buildings on rock cliffs above the beach east of the cove mark access to the mines. The lower mine just above the ruins was the Dickey Copper Co. Claim. which produced some 29,346 pounds of copper. In addition to the predominant minerals chalcopyrite, pyrite, sphalerite and pyrrhotite, minor amounts of lead , cobalt and silver are present. Above the Dickey property, between the 680 and 990 foot levels, were the more important Schlosser claims which were connected with the buildings on the beach by a 2700 foot aerial tram. The Schlosser claims recorded a long period of production from 1907 to 1920. Aside from Ellamar, the Schlosser was the biggest producer in the Port Fidalgo area accounting for 4,160,820 pounds of copper and 1,384 ounces of silver. The ore was mined from massive sulfide lenses and stockworks. Like the Fidalgo claim described above, the Schlosser is thought to still have a moderate mineral potential.

65. Porcupine Point, the western entrance point to Snug Corner Cove, displays turbidite features typical of middle fan deposits such as rip-up clasts in the lower sandstone layers, load casts and graded bedding. Foraminifera microfossils of a Paleocene-Eocene age were found on the west side of the point.

66. The Sheep Bay Pluton (Fig-35) is accessible from both Port Gravina and Sheep Bay. The Pluton has been dated to be about 52 million years old and seems to be associated with the Baranov-Sanak group of anatectic plutons which some

Fig-135. The 300 foot deep conglomerate beds at Simpson Bay are thought to be primarily inner fan deposits because of the greater proportion of pebbles and boulders compared to the finer grained matrix. Unlike the inner fan deposits at Galena bay, the Simpson Bay pebbles are well-rounded and consist of a feldspar, quartzite, tuff, granite and sandstone.

researchers think may have been caused by an interaction of the Kula-Pacific ridge with the Border Ranges Trench during Eocene times (Moore et al,1983). The pluton is exposed for over 55 square miles and is composed of a light gray granite with uniformly large crystals giving it a much coarser texture than the Oligocene plutons of the Sound. And unlike them it does not seem to exhibit a mafic phase. Inclusions of country rock up to 300 foot in diameter have been discovered in the granite. In many places contact with the turbidites has created a crushed silicified zone shot through with quartz veinlets. Because it is truncated on its northern end by the Gravina Fault, which is thought to be a branch of the Contact Fault System, the Sheep Bay Pluton has been cited as evidence that the Chugach and Prince William Terrane had to have been joined together by at least 52 million years ago.

67. Comfort Cove is a convenient place to observe the contact zone of the Sheep Bay Pluton. The north shore of the cove exhibits outcrops of shiney, baked metasedimentary rock having a slatey cleavage. Inclusions of country rock can be observed in the granite boulders lining the shore. These boulders offer fine examples of the coarse crystals of the Sheep Bay Pluton.

68. From **"Hole in the Wall Anchorage"** on Bomb Point one can explore the conglomerate beds of Simpson Bay. The best place to observe these rocks is on

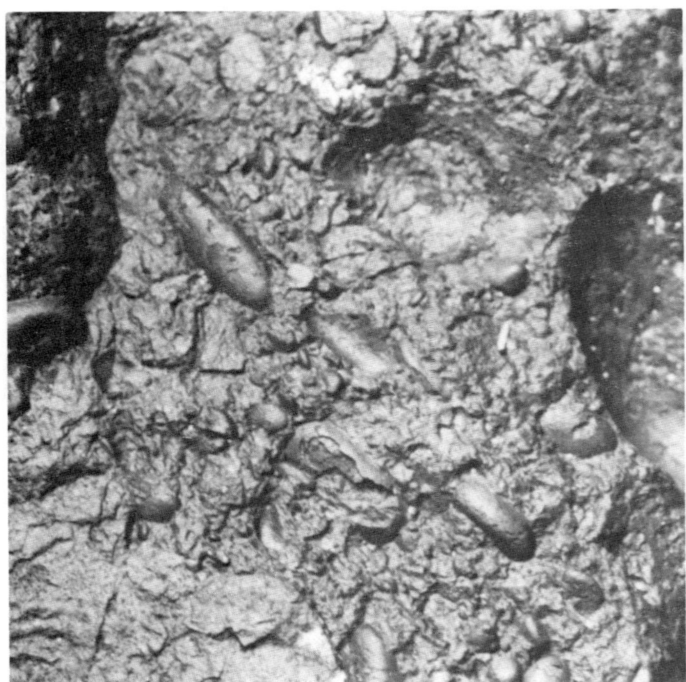

Fig-136. Careful examination of the Simpson Bay con- glomerates reveals places where the alignment of the pebbles reveals the direction of flow of the turbidity current which laid down these deposits.

the islands just north of the anchorage and the point behind them which divides the two arms of **Simpson Bay**. Outcrops of the same beds can be seen on Sheep Point to the west and at the head of the south arm. These conglomerates have been estimated to be over 300 feet thick. Like the Galena Bay conglomerates, the Simpson Bay rocks are thought to represent inner fan deposits and possibly the entry point of feeder channels to the middle fan (Winkler and Tysdal, 1977). However, unlike the Galena Bay deposits, the clasts are more rounded and consist of a coarse-grained feldspar, quartzite, tuff, granite and sandstone; whereas, the Galena Bay deposits consist of greenstone, argillite, sandstone and limestone clasts. Half of the rounded pebbles, cobbles, and boulders of coarse-grained of the Simpson deposits are rocks foreign to this area. A sandstone boulder 30 ft in diameter embedded in the sandstone matrix can be seen high on the south side of the point.

Bomb Point to the west of the anchorage exhibits some good folds typical of Orca Group rocks. Limestone concretions can be seen in the sandstone just to the east of these folds.

69. The town of **Cordova** escaped loss of life from the 1964 earthquake. More damage was done to the town by the six feet of uplift than by the shaking or tsunamis. Water in Orca Inlet did rise and fall a number of times but the uplifted town sustained little damage. However, swift currents in the inlet damaged some pilings and moored boats. The most significant impact was uplift of cannery and city docks; the small boat harbor had to be dredged before it could be used again; and the channel leading to the fishing grounds on the Copper River Delta had to be redredged.

Eyak ridge above Cordova consists of a turbidite sequence of sandstone, siltstone, and argillite interlayered with mafic volcanic rocks (pillow basalts and massive flows). A gravimetric high, extending from here across Hawkins and Hinchinbrook Islands, coincides with outcrops of basalt on the surface. That this high is weaker than the high extending through Knight Island and across the center of the Sound suggests that these mafic layers are probably not as thick and may have a different origin. Also, the association of limestone accretionary layers in the turbidites above these basalts suggest that they may have been extruded in shallower water than the lavas of the Knight Island ophiolite where no such limestones occur. Limestone dissolves or does not form at depths greater than 12,000 feet. This limestone sometimes contains microfossils of turbidite origin (i.e. shells transported from shallower depths by turbidity currents). Three foraminifera found in these limestones near Cordova indicate a late Paleocene to early Eocene age.

A good outcrop of folded Orca turbidites can be seen at the Cordova Ferry terminal and a conglomerate containing granite clasts can be seen directly across from the town on the south shore of Hawkins Island.

70. Point Whitshed south of Cordova is a good place to observe the Cordova-Hawkins-Hinchinbrook Island volcanic rocks. Mummy Island off the end of the point is composed of the same basalt.

71. The east shore of **Canoe Passage** is a good place to observe folded Orca turbidites, thin beds of dark impure limestone, and an outcrop of Hawkins Island basalt. In nearby **Cedar Bay,** a 1-2 ft. thick layer of a dark blue to black limestone can be observed.

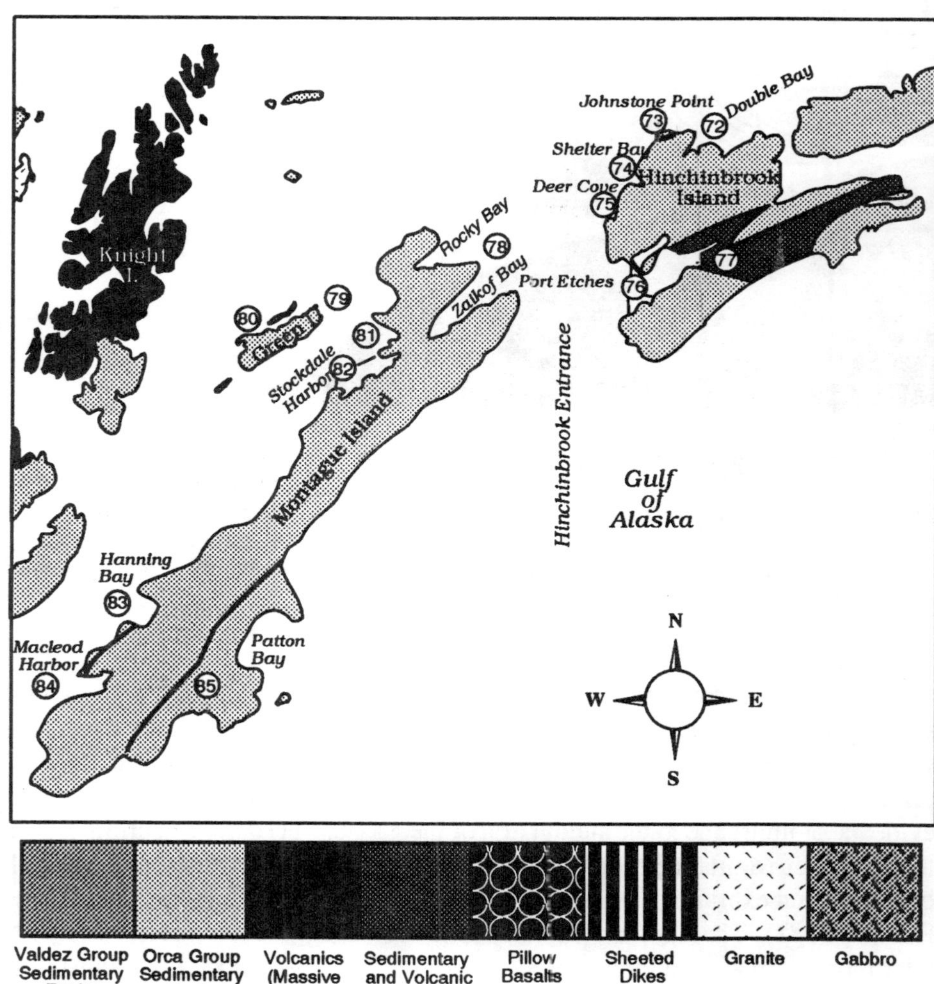

Fig-137. Geological Map of the Montague Belt Area.

VII. Montague Belt

General:

Montague, Hinchinbrook and Green Islands comprise a distinct set of rocks known as the Montague Belt (Helwig and Emmet, 1981). These rocks consist mainly of middle fan turbidites probably laid down in a westward sloping fan. What distinguishes these rocks from the rest of the Orca Group is their rare

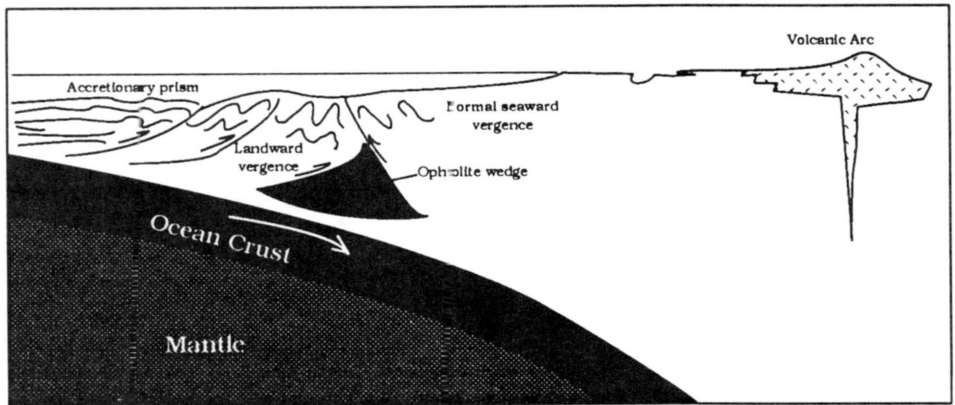

Fig-138. Rocks of the Montague Belt exhibit rare landward vergence in their isoclinal folds, which are inclined to the northwest rather than to seaward (all of the Prince William Sound region except the Montague Belt) as would be normal for a subduction trench. Helwig and Emmet suggest that the landward inclination may be a result of the rocks being thrust against the ophiolite (Knight Island) in the accretionary prism during subduction. Adapted from Helwig and Emmet (1981).

landward vergent structures. Isoclinal folds, apparently developed in these rocks shortly after deposition, are inclined to the northwest rather than to seaward as would be normal for a subduction complex. These folds were themselves overprinted by westerly to northwesterly oriented folds of Miocene age suggesting evidence of uplift due to the subduct on of the Yakutat Terrane. Superimposed on these folds on Montague Island are neotectonic structures due to more recent earthquakes which exhibit normal southeast (seaward) vergent structures (Helwig and Emmet, 1981). The anomalous, early landward vergent structures followed by later more normal structures may be evidence that these rocks were transported here from elsewhere and may have developed their early orientation in association with the Border Ranges Trench rather than the present Aleutian Trench. However, Orca rocks north of the Knight Island Ophiolite display normal seaward vergence suggesting to Helwig and Emmet that rocks of the Montague Belt may have acquired their present orientation by being thrust against the Ophiolite in the accretionary prism during subduction.

A greenstone belt extending from the head of Port Etches to the southeast tip of Hinchinbrook Island seems to be related to the outcrops of volcanic rock on Hawkins Island and near Cordova. The gravity and magnetic highs characteristic of these basalts are much weaker than those associated with the Knight Island ophiolite suggesting thinner sections perhaps having a different origin.

Microfossils recovered from thin beds of limestone and from limestone concretions on Montague and Hinchinbrook Islands yield Paleocene/Eocene dates for the rocks of this belt. Some of these broken, fossiliferous limestones seem to have originated in shallower water and been swept down into the subduction trench by turbidity currents; others may have formed in shallower waters on pillow basalt flows that reaches upward toward the sea's surface.

Montague Island is one of the best places to witness evidence of the 1964 earthquake; for it is here that the earth ruptured forming numerous thrust faults. Raised beaches line the shores. The southwest corner of the island was lifted as much as 35 feet.

72. The strata in and around the anchorage at **Double Bay** abounds in turbidite features such as: rip-up, ripple marks, flute, groove and load casts.

73. Poorly preserved conifer fossils and foraminifera microfossils have been observed on **Johnstone Point.**

74. Age diagnostic microfossils have been found near **Shelter Bay.** Look for sole markings and other turbidite features here.

75. Just north of **Deer Cove,** there is an outcrop of middle fan conglomerates.

76. Port Etches exhibits outcrops of conglomerate on the northwest shore of Constantine Harbor, on the west shore and head of English Bay, and along the western shore of the bay northeast of English Bay.

77. Garden Cove and the head of Port Etches are good place to observe the contact between the Hinchinbrook greenstones and Orca turbidites. Thin lenticular layers of limestone overlying the pillow basalts in this area have yielded shallow water upper-Paleocene microfossils. Fine mudstones overlying these limestone layers suggest that basalt flows may have overtopped an area of turbidity flow and that the limestones formed over local basalt highs (Winkler, 1976). Helwig and Emmet suggest that the source of these basalts may have been a leaky transform fault near a continental margin. Another possibility is that the shallow water limestones were deposited in deeper water by turbidity currents.

78. Outcrops along the shores of **Zaikov Harbor and Rocky Bay** have yielded poorly preserved carbonized plant remains including carbonized alder pollen of

Tertiary age. Borings and feeding trails of marine animals can be found in calcareous and concretionary layers in this area. Ripple marks and other sole markings occur in outcrops along the beaches.

79. The northeastern tip of **Green Island** can be reached from the northern end of Gibbon Anchorage. A good display of load casts occurs on the southern shore of the long island protecting the anchorage about one mile from its northeast end. Outcrops along the shore of the northeast tip of Green Island contain numerous excellent examples of turbidite features including ripple marks, load casts, rip-up, flute and groove casts. There is a ripple marked sandstone face on the north shore of this point about 3/4 miles from its tip. Numerous calcareous concretions can be found in the slate beds of the beaches in this area. Iron rich, metallic nodules are present in the graywacke at several locations. To the south of the point, a fine-grained conglomerate grades into a gritty then a coarse grade sandstone. An outcrop of sandstone just south of the extreme N. E. tip of the island exhibits a bed of fragmented carbonized plant remains. Frondescent load casts can also be seen in this outcrop. Just south and east of here a small outcrop on the beach reveals pronounced ripple marks on a sandstone face. Calcareous concretionary layers and

Fig-139. Praying Hands Rock was first worn by tidal currents than uplifted.

Fig-140. Fine-grained conglomerates occur in a bed on the NE side of Green Island.

individual concretions (cf. p. 115) occur in mudstone outcrops along the beach.

80. One can walk the beach from western **Gibbon Anchorage** around Putnam Point to the southwest end of Green I. Several outcrops of middle fan turbidites uplifted to the vertical can be examined along the beach. These consist of alternating beds of mudstone and sandstone with occasional conglomerate. The conglomerate clasts are rounded pebbles of quartz, greenstone, granite and mudstones.

In many places the mudstone beds have eroded away leaving exposed faces of sandstone on which flute casts (cf. p. 112) and load casts (cf. p. 115) can be observed. In places thin beds of reddish iron rich (hemipelagic) slate alternate with thin sandstone beds. Some of the sandstone beds contain calcareous concretions and occasional metallic concretions (Manganese/iron?) exhibiting an intense iron staining.

81. Calcareous concretions up to 2 feet in diameter occur in sandy beds on the south side of **Stockdale Harbor**.

82. Point Gilmore exhibits a conglomerate composed of well-rounded sand-

Fig-141. Iron stained concretions occur in sandstone beds on SW Green Island.

Fig-142. Flute casts found on a sandstone boulder on the SW side of Green Island.

stone and mudstone clasts. The point 4 miles to the southwest of Point Gilmore has a conglomerate bed of a coarse sandstone matrix containing small, well-rounded clasts of quartz, granite, and diorite.

83. Flute, groove, and load casts can be observed on the northern entrance point to **Hanning Bay**.

The Hanning Bay fault which ruptured during the 1964 earthquake can be traced from the southern entrance point of Hanning Bay across the small drying Bay to the south and to the northern entrance point of Macleod Harbor. About 16 feet of vertical movement occurred along this four-mile-long reverse fault in 1964.

84. Conglomerates outcrop on the northern entrance point to **Macleod Harbor.** Maximum uplift of Montague Island occurred here during the 1964 earthquake considerably reducing the extent of this bay.

85. A maximum displacement of 36 feet occurred along the 22 mile long **Patton Bay Fault** during the 1964 earthquake. The spectacular fault scarp with the northwest side thrust above the southeast side is still visible today.

Fig-143. The northwest side of the Patton Bay Fault is thrust above the southeast side.

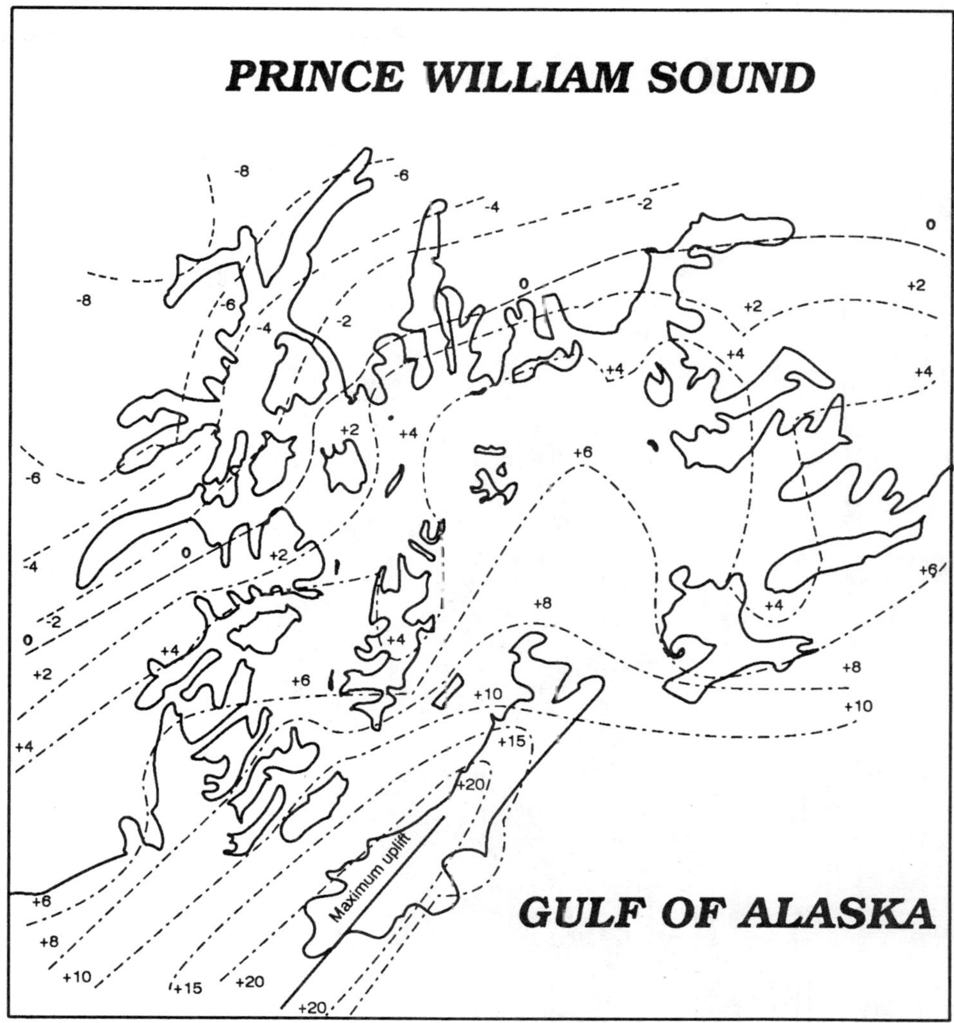

Fig-144. Map showing changes in the land as a result of the 1964 Earthquake. A "--- ---" indicates a contour showing the subsidence in feet. A "— . — " line indicatges a contour showing uplift in feet. Adapted from "Areas of Tectonic Land level Changes . . . During the 1964 Earthquake," D. Blanchet, *Chugach National Forest Environmental Atlas.* USDA-FS, FS Alaska Region Report Number 124.

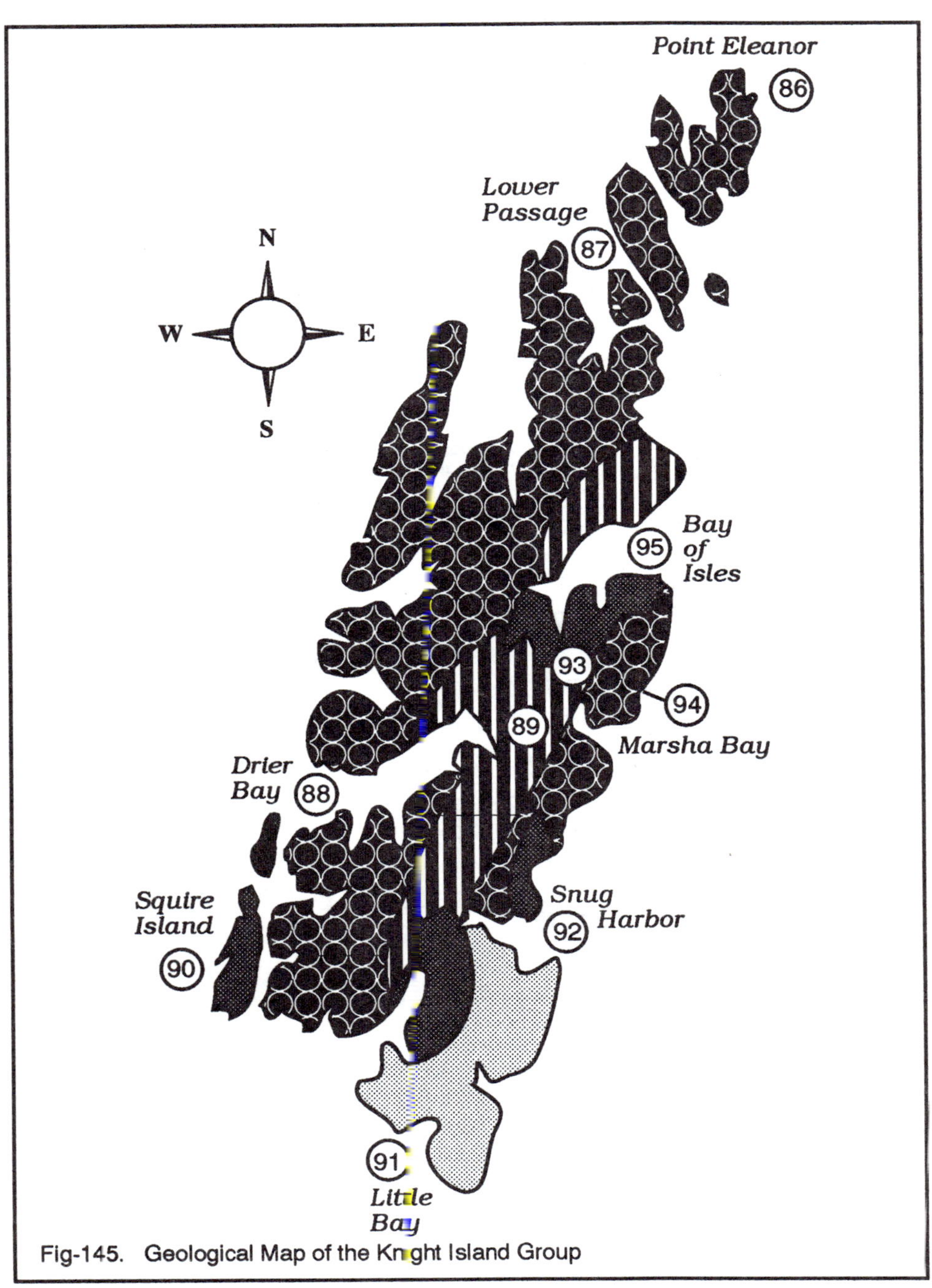

Fig-145.   Geological Map of the Knight Island Group

VIII. The Knight Island Group

General

Knight Island is one of the most geologically interesting areas in Prince William Sound. The Knight Island group seems to represent a slice of ocean crust ("an ophiolite") that originated at an oceanic spreading center perhaps about 57 million years ago. These pillow basalts and underlying sheeted dikes were later transferred across a subduction zone (i.e.obducted) onto the crust of the North American continent. Unlike most ophiolites, the Knight Island sequence lacks an overlying layer of pelagic sedimentary rock (chert) created by a continuous rain of the silica-rich bodies of deep water microorganisms. Secondly, layered gabbro underlying the sheeted dikes have not been exposed. However, small gabbro intrusions and inclusions of gabbro and peridotite occur in several locations indicating that these rocks may underlie the exposed greenstones. Gravity and magnetic measurements suggest the presence of mafic and ultramafic material beneath the surface. A six-mile thick layer of dense ultramafic material is necessary to explain the high gravity measurements. Finally, the pillow basalts near the Bay of Isles and on the southern end of the island are interbedded with ashy layers of typical Orca turbidites. The lack of deep-water sediments and the interbedding of the volcanic rocks with turbidites suggest that these rocks originated at a spreading center located near a continental margin.

The Knight Island Ophiolite, then, may representsa chunk of the Kula-Farallon Ridge which was interacting with the Border Ranges Trench as the rocks of the Chugach/Prince William Terrane were being carried northward aboard the Kula Plate sometime during the early Eocene.

Turbidity flows cascading down into the trench buried the sporadic extrusions of lava oozing from the spreading center creating interlayered turbidites and pillow basalts. The Ophiolite, thus, was probably incorporated into an accretionary prism which was later offscrapped onto the North American Plate. During uplift (probably during the Miocene – about 15 million years ago) the Knight Island rocks were bowed up into a broad anticline and regionally metamorphosed into green-schists (greenstones). The western limb of the anticline facing Knight Island Passage dips gently westward exposing a three mile thick layer of pillow basalts whose tops face westward. The eastern flank of the island consists of a steeper dipping but thinner layer of pillow basalts whose tops face eastward. Thrust upward between these opposed pillows in the central core of the island are the sheeted dikes which once acted as underlying volcanic feeder vents to the extruding pillows.

The western and northern shores of the island group offer the most abundant displays of undeformed pillows. Most are composed of a dark to fine grained basalt with larger crystals forming in the interiors of some of the larger pillows. Some of these exhibit radial cooling cracks which are sometimes filled with epidote veins and contain vesicles formerly occupied by gas bubbles but now filled with chlorite. One geologist has measured these vesicles to calculate the pressure created by the overlying waters and hence the depth of water covering these rocks during their formation. He estimates that the pillows were formed at depths of from 1,000 to 1,700 feet (McGlasson, 1976). Although most pillows range from one to three feet in diameter, some approach 12 feet. Some are multilobed. The fractured surface of many of the pillows and the shattered breccia between pillows indicate rapid cooling. Red and green chert formed from the interaction of the hot lava with sea water fill the cracks between some pillows.

Sheeted dikes in the central part of the island can best be observed at the Bay of Isles and at the heads of Snug Harbor and Drier Bay. Feeder vents for the pillows are usually vertical, exhibit a north-south trend, and vary in size from a few inches to 100 feet wide but are normally about 4 to 5 feet wide. Their sheeted form and sometimes chilled margins suggest repeated intrusions of older by younger dikes. The dike material is usually basalt but occasionally quartz, gabbro or diorite.

Ophiolites in other parts of the world such as Cyprus and Japan have yielded major copper discoveries. Although the concepts of ophiolites and black smokers were unavailable to the early prospectors of Prince William Sound, they were keenly aware that Knight Island's volcanic character held great promise for mineral exploration. Between 1903 and 1909, virtually the entire island had been staked with claims; in all, 42 prospects were on file by the end of this period. However, Knight Island has failed to live up fully its promise; only two small mines in Drier Bay and the Pandora Mine ever shipped any ore. These were only trial shipments and probably only amounted to a few thousand pounds.

Unlike the copper deposits of the Midas, Ellamar, Port Fidalgo and Latouche mines, which occur in sedimentary rocks associated with nearby greenstones, the numerous, small Knight Island deposits are hosted almost exclusively in volcanic rocks. The metallic sulfides occur mainly as disseminated grains in and between basalt pillows and in shear zones in the sheeted dikes unit. The main copper sulfides are cubanite and chalcopyrite. One prospect on Knight Island has continued to attract interest – the property at Rua Cove. Core sample were taken from this site in 1972-73 and again in 1976. The volcanic rocks in this area seem to be cut by a network of distributary feeder vents to a former submarine hot springs. Although the quantity of ore seems sizable at this location, it is not concentrated; and the Bureau of Mines rates it as having only a moderate mineral potential.

86. Point Eleanor, near the navigation marker, has a good outcrop of pillow basalts observable from a boat.

87. Lower Passage is a good place to observe geological features typical of the Knight Island group. From the protected Disk Island anchorage, one can explore the island's southern shore. This is an excellent location to observe typical pillow basalt features including pillow breccia, multilobed pillows, radial cooling cracks and vessicles. The cove about 1\2 mile south of the entrance to the anchorage contains a 10 foot adit near its head on the east shore. On the east side of the eastern entrance point to this cove, one can find some intensely malachite stained pillows. The easternmost point of the island displays the characteristic iron-stained pillows which mark most mineralized areas of the Knight Island group.

The east arm of Louis Bay to the south of Disk Island preserves traces of the area's mining history. Two prospects have been dug into the west shore near the head; near here, the remains of a miner's cabin marks the trail head to the main prospects at 550 ft. and 900 ft. above the bay. The pilings on the southeastern shore of the bay mark the location of a sawmill and steam plant dating from 1908.

88. Drier Bay hosts a number of mines and prospects and provides access to geological features characteristic of the rugged central section of Knight Island. The greenstones along the east and southeast shores of the bay contain xenoliths of gabbro and peridotite. Look for light greenish gray to dark inclusions in the massive greenstones.

Fig-146. The rind of a pillow basalt on Disk Island, Lower Passage.

Fig-147. A basaltic pillow intruding a tuffaceous layer, Squire Island Anchorage.

An old mine tunnel in iron-stained, mineralized rock can be explored at the south end of Cathead Bay while several tunnels penetrate the cliffs between Mallard and Barnes Cove. At Barnes Cove old mining wreckage can be seen on the beach presumably supporting the claims of the Copper Coins Group 1/2 mile southwest of the cove at the 518 ft level or the Hercules prospect 1/2 mile southeast of the cove between the 300 ft and 1000 ft. levels.

Another mine is located at the 700 ft. level 1/2 mile southwest of the head of Northeast Cove. The mountains at the head of Northeast Cove are a good place to observe sheeted dikes. A zone of sheared rock separating the pillow basalts from the zone of sheeted dikes (the Port Audrey Shear Zone) extends through Northeast Cove to the head of Mummy Bay. One mile south of the head of the cove in this shear zone is a body of dense peridotite 650-350 ft. in diameter – a possible clue to what may lie below.

The two most successful mines in Drier Bay were above Port Audrey: the Jonsey prospect and the Nellie prospect. The Jonsey prospect is located in a mineralized shear zone containing pyrrhotite and chalcopyrite 1000 feet up on the ridge 1/2 mile east of the lagoon. Cables from the tramline to this mine can be found on the shores of the lagoon. The mine shipped only a small quantity of ore. The Nellie group is in a steep walled ravine 3/4 miles east of Port Audrey and also contains pyrrhotite and chalcopyrite.

89. The sheeted dikes to the east of Port Audrey contain small intrusive bodies of gabbro and granodiorite. The former may offer a clue as to what lies below the sheeted dike complex if Knight Island is indeed a true ophiolite, while the latter is probably a differentiate from the mafic rocks.

90. Squire Island Anchorage is a good place to observe the interbedding of greenstones and turbidites. At its head in the eastern bight, one can see lavas intermingling with sandstones and slates. In the western bight, pillows intrude thick, tuffaceous layers of sediments.

91. Possible slump deposited conglomerates lie from **Little** to **Mummy Bay.**

92. Discovery Point on the south side of Snug Harbor exhibits turbidites typical of the southern end of Knight Island. Here, intricately folded thin beds of a slightly calcareous dark-gray slate alternate with a lighter colored but coarser grained slate. Calcareous nodules up to 15 inches in diameter appear in the slate. There are a few thin beds of limestone. The northern entrance point to the bay has an outcropping of conglomerates.

93. The Pandora Group of mines can be reached either from the southern arm of Bay of Isles or from the head of Marsha Bay. The mines are located 1/2 mile southwest of the south arm of the Bay of Isles. The workings lie about 400 feet above sea level on the west bank of the main stream flowing into head of the arm. The shaft intersects a 5 ft. lens of good sulfide ore in greenstone. The mine was one of the few on Knight Island that ever shipped any ore.

94. The Rua Cove Mine represents the largest established copper reserves on Knight Island. The mine tailings can be seen from a distance off shore but not from small, unprotected Rua Cove. A trail leads from the buildings on the shore of the cove to the mines.

The deposits were discovered and claimed by Charles Rua in 1905. By 1908, Charles Rua and J.J. Bettles had begun work on the property. Some ore was shipped that year for testing purposes. In 1912, Fred Liljegren filed Copper Bullion claims #1 and #2 on the property. W. A. Dickey of Seattle acquired the claims in 1916 building a water power plant and several buildings on shore. With the aid of a compressor, he completed 1750 feet of workings. In 1929, a Canadian Company purchased the property, made some improvements and extended the workings. However, they dropped their option during the depression in 1930. Further studies of the claims were conducted in 1938, 1942, 1943, and 1948-49. Further cores were drilled on the property in 1964, 1972-73 and in 1976. Most of the work has been done in a shear zone which extends from the bottommost tailing pile on up the mountain. Sulfide ore has been exposed at the 550, 620, 675, 705, 1000 and 1250 foot levels. Diamond drilling over a wide area exposes what seems to be a stockwork deposit of vents that once fed a submarine hot springs. There are probably several million tons of low grade .6% copper ore at this site. In the sixties, a proposal was published suggesting that these low-grade deposits be mined with a nuclear blast (Williams, 1966.)

95. On the South Shore of The Bay of Isles, sedimentary beds alternate with lava flows. Greenstones on the north shore are intruded by magnetite bearing gabbros. In some intrusions magnetite makes up 15% of the volume of the rock. Mariner's magnetic compasses are known to be inaccurate in this area because of a strong magnetic anomaly. This strongly magnetic gabbro may be representative of the rocks underlying the island group which exhibits a strong aeromagnetic signature suggesting that Knight Island represents a true ophiolite sequence.

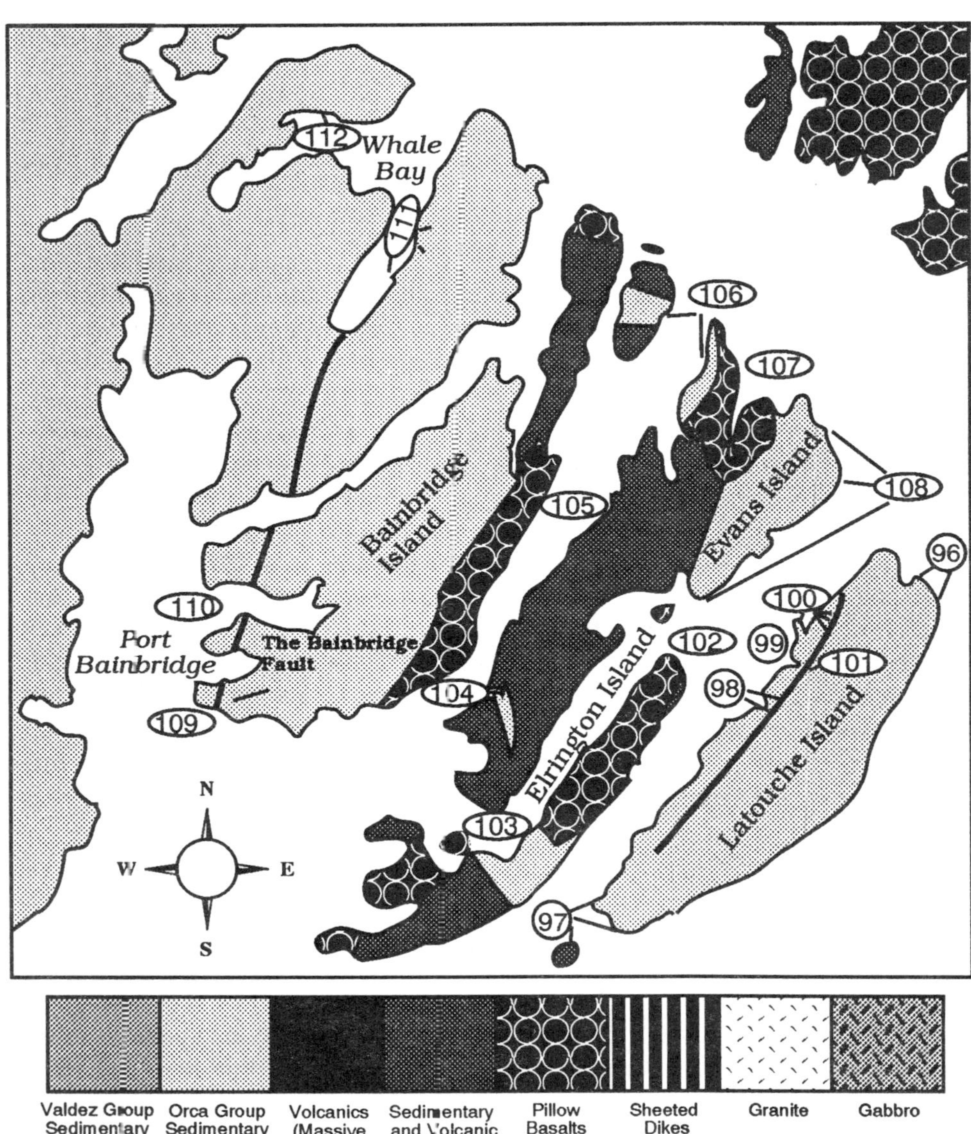

Fig-148. Geological Map of the Southwest Islands Group

IX. Southwest Islands

General

Helwig and Emmit (1981) classify the rocks of the southwestern corner of the Sound as belonging to three more or less distinct belts — the Montague Belt, the Bainbridge Melange, and the Blying Sound Belt. The rocks of Latouche Island consist of a sequence of Orca turbidites in many ways resembling the rocks of Montague Island and the southern end of Knight Island. Here, slate or argillite beds alternate repetitively with layers of graywacke, which are occasionally underlain by conglomerates; a few thin layers of limestone are also in evidence. Like the Montague rocks, folds in these rocks are landward vergent. However, the Latouche rocks are more highly metamorphosed than those on Montague Island and are intruded by mafic rocks (on both its northern and southern ends.) Also the rocks of Latouche Island host the richest metal sulfide deposits in the Sound; whereas little mineralization has been discovered on Montague Island.

The Latouche Island rocks seem to be transitional between the rocks of the Montague Belt and those of the Bainbridge Melange that outcrop on Elrington, Evans and Bainbridge Islands just to the northwest. The intensely deformed rocks of the Bainbridge Melange lie east of the Bainbridge Fault and consist of disrupted blocks of basalt intruded into Orca sediments. These appear to grade from mixed volcanic and sedimentary rocks on the east into the purely mafic rocks of the Knight Island ophiolite on the west. Extruded pillow basalts (often mixed with sediments on Bainbridge Island) and unintruded sediments also occur on these islands. Minor beds of conglomerate, limestone and chert also occur here.

Rocks to the west of the Bainbridge Fault and along the western shore of Port Bainbridge appear to be quite distinct from the above two groups. Here, massive layers of sandstone replace the frequent bedding planes of the Montague Belt; and the intensely folded, turbidite beds show normal southeasterly (seaward) vergence rather landward vergence. Because this band of massive sandstones extends from Blying Sound on the south through Icy Bay and Nassau Fiord to Unakwik Inlet on the north, Helwig and Emmet refer to these rocks as "the Blying Sound Belt." Rather than being intruded by the basalts of the Bainbridge Melange, these rocks are intruded by the Sound's Oligocene plutons.

96. A 100-foot-wide gabbro dike cuts the turbidite sequence on the southern entrance point to **Sleepy Bay**. About one mile south of this location, on the southeastern tip of Latouche Island is an outcrop of conglomerates.

97. Mafic volcanic sills outcrop on Danger Island and on the southwest point of **Latouche Island**. A 12-15 foot bed of conglomerates can be found on the northwest point of Latouche Island at Pleasant Bay. The conglomerates underlying graded graywacke and slate beds are typical of those found elsewhere on the island. Clasts in the conglomerate are small, well rounded, pebbles of granite, quartz, quartzite, chert, argillite and slate.

98. The shoreline around **Horseshoe Bay** offers opportunities for viewing turbidite features such as graded bedding, conglomerates, sole markings and rip-up. The anchorage also offers access to the Duke and Duchess mines. To reach the mines, follow the clearings containing the remains of the former mine buildings on the southwest side of the stream. Edge over toward the stream until you find the old, rotten, corduroy road leading to the mines. The Duke mine lies on the western side of the stream and is presently flooded; but to reach the more interesting Duchess Mine, the stream must be forded where the road ends. Ascend the meadows into the next valley to reach the mine. A building containing corings stands beneath the two large tailing piles accumulated from adits at the 400 and 500 foot levels. Boulder-sized samples of high grade pyrite can be found in the tailing piles.

The Duke and Duchess Mines were staked in 1898 and were owned by the Reynolds-Alaska Development Company. Work began on the properties between 1905 and 1907 and continued to 1916. In 1906, the Duchess mine produced some 215,000 lbs. of copper. By 1907, the Reynolds company constructed a trading store, a light plant and a number of houses on the shores of Horseshoe Bay and finished the mile-long corduroy road to the mines in 1908. By 1916, when the mines were closed, there were over 2,800 ft of workings at the Duchess site but no further shipments of ore were made. The main ore body of the Duchess was a 490 foot massive sulfide lens consisting of almost 100% pyrite. The deposit lies along the Beatson Fault and and is hosted in slate and graywacke. During 1952 and again in 1955, during a worldwide sulfur shortage, the pyrite was considered as a possible source of sulfur for the manufacture of sulfuric acid. The corings in the shack below the tailing piles were probably left by the Northwest Exploration Company which again investigated the deposit in 1970.

99. Conglomerates outcrop at **Wilson Bay**. These may be continuous with the deposits at Pleasant Bay and Powder Point.

100. Latouche near Powder Point was the major copper mining center in Prince William Sound. There were actually three mines at this location: the Beatson (Beatson-Bonanza), the Chenega, and the Ladysmith (or Girdwood) mines. The Beatson mine, located 0.5 miles southwest of Powder Point was the largest

Fig-149. Top. Early ore bunker at the Beatson Mine on Latouche Island. Photo by Sidney Paige, USGS, 1905. USGS Photographic Library. Fig-150. Bottom. The mining town at Latouche sometime after 1917 when the bunkhouse was constructed. This photo is taken from near the same spot as Fig-149. Several of the early buildings are still standing. Panoramic View Collection. Alaska Historical Library.

producer of copper in the Sound and second only to the Kennecott mines in the Wrangell Mountains as the largest producer in Alaska. During its lifetime, this mine produced 182,600,000 lbs. of copper, 484 oz. of gold, and 1,466,649 oz. of silver.

Discovered in 1897, the Beatson shipped its first ore in 1899 and continued with regular shipments to the smelter in Tacoma from 1904 until it closed down in 1930 when its reserves were considered to be exhausted. Although considerable quantities of high grade ore were taken from the property during its early days, open pit mining methods made it economically feasible to continue operations as long as ore averaging 1.5% copper remained. Much of the ore was concentrated in mills on location before shipping to the smelter. The Beatson was the last copper mine in Prince William Sound to close.

The property was acquired by Kennecott in 1910 and the small town of Latouche grew up near the mines. By 1916, Latouche boasted a population of over 400, had a Post Office, hospital, electrical generation plant and wireless station and became a major supply port for the district with regular steamer service to Seattle, Valdez and Cordova. The old mining town still stood on this site in the early 1970s until it was bulldozed by an Anchorage developer to make room for the present housing development.

The ore was principally chalcopyrite, pyrrhotite and pyrite hosted in slate and mainly graywacke. The main ore body was a large massive sulfide lens 1000 feet long and 400 feet wide. Disseminated sulfides were also mined. The ore bodies of all three mines in the area follow the trend and seem to be associated with the Beatson Fault as are the Duke and Duchess mines farther to the south.

A 75-foot thick bed of conglomerates outcrops on **Powder Point**. The matrix is a coarse-grained graywacke with well rounded pebbles and cobbles up to one foot in diameter. The clasts appear to be quartz, granite, quartzite, slate and some coarse-grained, mafic rocks.

101. The Beatson Fault, which runs almost the entire length of Latouche Island, dips steeply to the west and is marked by a zone of gouge, sheared and shattered rock, and cavey ground. Most of the important sulfide deposits of the island are located along this fault probably because these ore bearing rocks, which were most likely deposited by black smokers in submarine hot springs, are less competent than the surrounding turbidites.

102. Elrington Island exhibits considerably more greenstones than outcrop on Latouche Island. Elrington Island marks the southwest end of the gravity high that extends northward through Knight Island to Glacier Island and then eastward. These mafic rocks occur as extruded, pillow basalts on both the southern and northern ends of the island and as intruded sills adjacent to the extrusions.

103. Conglomerates similar to those on Latouche and Evans Island can be viewed just west of the spit and island at **Fox Farm Anchorage**.

104. Gabbro intrudes the Orca sediments at this location; look for quartz extensions of the gabbro body and boudinaged quartz veins in the slate.

105. Disrupted mafic blocks, typical of the Bainbridge Melange, can be seen along the southern shores of **Evans Island**.

106. Lenses of limestone about 6 feet thick and 30 feet long outcrop on **Flemming and Evans Islands**. A sea lilly fossil (B.L. Johnson, 1918) and more recent a microfossil (foraminifera) were found at this location on Evans Island.

107. A 15-20 foot thick band of chert weathered white lies on the southwest shore near the head of **Shelter Bay**.

108. Conglomerates, probably middle fan deposits, similar to those on Latouche Island outcrop at these locations.

109. The **Bainbridge Fault** can be viewed crossing Point Pike. This west dipping thrust fault juxtaposes the massive sandstones of the Blying Sound Belt over the slatey and intensely deformed sediments of the Bainbridge melange.

110. The **Hogg Bay conglomerates** are similar to those found on Evans and Latouche but contain more greenstone clasts. The proximity of these conglomerates to the Bainbridge Fault has resulted in clasts which have been sheared and deformed.

111. The Gold mine on the northern entrance point to **Minke Cove** in Whale Bay is one of several exceptions (cf. also Jackpot Bay, Bligh Island and Culross Island) to the notion that Gold deposits occur only in Valdez group rocks while copper deposits occur in Orca Rocks. Conglomerates occur along the eastern shore of the eastern arm of Whale Bay.

112. The prominent cliff of pink sandstone in **Orca Cove** is typical of the massive sandstone beds of the Blying Sound Belt.

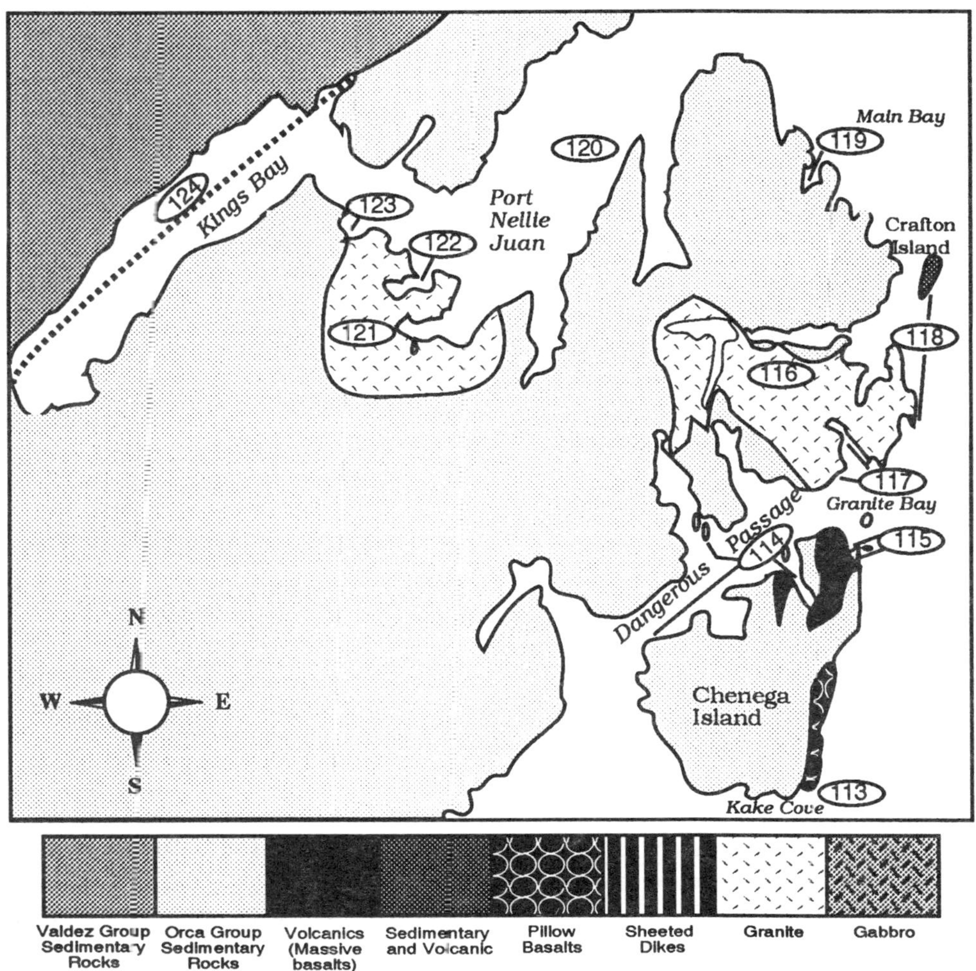

Fig-151. Geological Map of Rocks on the Western side of Prince William Sound

X. Western Sound

General

Except for the rocks of Chenega Island, which resemble the rocks of the Bainbridge Melange to the South, the rocks of this area seem to belong to the Blying Sound Belt of Helwig and Emmet. These turbidites are intruded by two Oligocene Plutons – one near Eshamy Bay and the other close by at the head of

Fig-152. Intensely folded rocks of the islands near the mouth of Ewan Bay.

Port Nellie Juan. Contact metamorphism of the sedimentary rocks between these two plutons and between them and the Culross Island Pluton suggest that they all may be connected at depth (Tysdal and Case, 1979).

113. Kake Cove. The eastern shore of Chenega Island has basalt pillows interbedded with a coarse tuff.

114. Dangerous Passage contains several geological features of note. Conglomerates are found along the western shore of Chenega Island just south of Masked Bay and near the island's southwestern tip. Volcanic rocks intrude the sedimentary rocks of Masked Bay on its western entrance point and at its head. The islands inside the mouth of Ewan Bay along its north eastern shore are formed from intensely folded sedimentary rock – probably due to their proximity to the nearby Eshamy Pluton.

115. Just inside the cove to the north of the offshore islets off the northeast tip of **Chenega Island** is an unexploited manganese deposit. The deposit is along the south shore in an outcrop of black rock that stands out from the shoreline cliffs. The manganese minerals occur in a black calcareous chert bed exhibiting a weathered iridescent surface. Freshly broken samples of this reveal pink deposits of the manganese minerals. About 1/4 miles south of this location in a small, shallow cove behind a hooked point and marked by prominent thinly bedded strata are some excellent examples of overlapping flute casts.

116. The **Eshamy Pluton** is accessible from the heads of Ewan and Paddy Bays or along the shores of Granite and Eshamy Bays. The Eshamy Pluton has been dated to be between 35 and 36 million years old. The medium-grained Eshamy granite resembles that of the Nellie Juan Pluton to the north but contains hornblende in addition to biotite as its darker minerals. These range from scattered dark specks to black blotches up to 1/4 inch in diameter. Freshly broken pieces of the granite are a pinkish-grey color weathering to a darker pink near the surface due to iron staining. Like the Port Wells and Esther Plutons, the Eshamy Bay body has a mafic phase along its marginal areas – the gabbro intruded by the granite. The contacts with sedimentary country rocks are sharp with aplite dikes often cutting the turbidites. The contact rocks in places intricately mingle granite and sedimentary rocks giving the whole an almost gneissic appearance.

117. Granite Bay is a good place to observe many of the above features. The Contact zone between country rock and the Eshamy Pluton is readily visible on the northwest shore of Dangerous Passage, one half mile south west of the mouth to Granite Bay. Here, a number of aplite dikes cut the country rock, and the granite exhibits numerous inclusions. In places, banded granite suggesting fractional differentiation and flow can be observed.

About 1/4 mile northeast of the contact, behind a small treed island about 50 ft offshore, a large block of granite has separated from the cliff. Inclusions of country rock in the granite show flow structures where the partially melted sedimentary rocks have been stretched to conform longitudinally with the flow of the granitic magma.

Farther within Granite Bay, a point of dark rock along the south shore and the dark tip of the island opposite are gabbro. The mafic and felsic phases of the intrusive body seem to have been mixed while still plastic – the more felsic phase seems to have intruded the sediments before the mafic phase.

118. Volcanic rocks similar to those of Knight Island outcrop on Crafton Island and at Point Nowell. The **Point Nowell** rocks are in contact with the granites of the Eshamy Pluton and appear to be intruded by them (Tysdal and Case, 1979). If these volcanic rocks are associated with those of the Knight Island Ophiolite across Knight Island Passage, we might conclude that the Ophiolite is at least older than 35 million years.

119. Conglomerates occur near the head of **Main Bay.**

120. The north shore of **Port Nellie Juan** is composed of slatey, metamorphosed turbidites. The South shore consists of of slatey beds mixed with less metamorphosed beds preserving original sedimentary, turbidite features such as sole markings. Rocks nearer the Nellie Juan Pluton have, of course, been altered by contact metamorphism.

121. The Nellie Juan Pluton is a roughly circular intrusion of granite lying at the head of Port Nellie Juan. It appears to have only recently been unroofed by glacial action. From the glacial moraine above Derickson Bay, one can observe a mountain whose core has been exposed by the bulldozing action of a glacier. Here, the contact zone between overlying country rock and underlying, intruded granite is clearly visible (Fig-53). The Nellie Juan granite varies in color from a light pink to light grey. Large gabbro boulders in the moraine beneath the glacier suggest that, like the Eshamy Pluton, the Nellie Juan Pluton may have a mafic phase.

The contact zone cuts through Nellie Juan Anchorage. Aplite dikes occur in sedimentary rocks along its eastern shore. Inclusions can be observed in the granite along the southern and western shores.

122. Deep Water Bay is an excellent place to examine erosional features of granite. An exfoliated granite dome, reminiscent of those found in Yosimite Valley, towers above Deep Water Bay. Exfoliation and cleavage along cooling cracks in the granite have resulted in numerous large granite blocks in the area. Along the bay's southern shore and especially along the ridge in the valley above the bay, the granite is of such a composition that it is easily eroded. Streams carrying the sands resulting from this erosion have created at the head of the bay a white-sand beach. Numerous inclusions of country rock stand out boldly and uncharacteristically from the more easily eroded granites along the southern shore.

123. Greystone Bay contains a dramatic cliff face of exfoliated granite.

124. Some geologists identify **Kings Bay** as being formed by the Contact Fault.

Fig-153. Top. The granite domes of Nelle Juan Pluton tower above Deep Water Bay.
Fig-154. Bottom. Boulders along the southern shore of Deep Water Bay contain inclusions of country rock which stand out uncharacteristically from the more easily eroded granite.

Glossary

Accretion: The addition of new crust to a continental margin through the offscraping of sediments at a subduction trench or through the "docking" of a terrane.

Accretionary wedge" or "Accretionary prism": Great piles of sediment that accumulate along a subduction zone. Some sediments are offscraped against the inner trench wall and lithified. Others are partially subducted as coherent strata; whereas others are offscraped, shortened, and underplated onto the continental crust forming a slate belt.

Aeromagnetic high: Regions above the Earth's crust which give anonymously high magnetic readings, thought to be caused by the presence magnetic minerals in mafic and ultramafic rocks.

Anatectic plutons: Plutons that form in an accretionary wedge through the melting of sediments or through the absorption of sediments by molten basalt (as opposed to plutons that form above a subduction zone over the zone of partial melting.)

Andesite: An explosive lava extruded along magmatic arcs, the extrusive equivalent of the granitic, intrusive rock diorite.

Aplite: An intrusive igneous rock often found forming dikes or sills in or near contact zones. Often recognizable by its sugary texture.

Argillite: A hard, weakly metamorphosed mudstone resembling shale; however, unlike shale, argillite shatters rather than cleaves when dealt a hammer blow.

Asthenosphere: The mobile portion of the upper mantle over which rigid lithospheric plates slide.

Basalt: A heavy, dark, fine-grained, mafic igneous rock extruded at mid:ocean spreading zones, consequently forming the upper ocean crust. Basalts in Prince William Sound are found as pillow flows, as massive flows and as sheeted dikes and are metamorphosed to a greenschist and are often referred to as "greenstones."

Batholith: A large pluton of regional extent.

Black smokers: Chimney-shaped under water hot springs near oceanic spreading zones which spew forth a black "smoke" rich in metallic sulfides.

Blueschists: Rocks metamorphosed deep in a subduction trench by high pressure and low temperature. They contain the bluish mineral glaucophane which forms only under these conditions. Blueschists like tectonic melanges provide clues to the existence of former subduction zones.

Border Ranges Trench: A fossilized subduction trench marked presently by the Border Ranges Fault which stretches in a an arc from Baranof Island in Southeast Alaska to Kodiak Island to the west. The Glenn Highway between Glennallen and Palmer follows the trace of this fault zone. The Border Ranges fault divides the Chugach/Prince William accretionary terrane from the rocks of the Superterrane. It is thought that the rocks of Prince William Sound were formed in this trench probably at a latitude of 40°.

Breccia (Broken rock): Rock composed of broken angular rock fragments usually cemented in a finer matrix. Shattered pillow lavas and rocks occupying shear zones occur as breccias in Prince William Sound.

Calcareous: Rocks rich in calcium carbonate: limestones.

Chert: A flinty rock, high in silica which fractures conchoidally. Used in making arrowheads. Formed as the upper layer of some ophiolites from a mid:ocean rain of the silica skeletons of micrc:organisms called "radiolaria." Also formed when molten, extruded basalts react chemically with seawater.

Chugach Terrane: An older tectonic melange bordered by younger sequence of trench-fill turbidites which stretches 1200 miles in an arc from Baranof Island in the east and Sanak Island to the west. The Chugach and the Kenai Mountains of the Prince William Sound area occupy the Chugach Terrane. This terrane is often defined as being bordered or the north by the Border Ranges Fault and on the South by the Contact Fault.

Clasts: Rock fragments or grains cemented together in a finer matrix in sedimentary rocks.

Concretions: Lumps or nodules usually of calcium carbonate enriched rock occurring in shales or sandstones: usually formed from the precipitation of calcium carbonate around a nucleus from concentrated mineral solutions during the formation of the host rock.

Conglomerate: A sedimentary rock composed of pebble, cobble and occasionally boulder:sized clasts embedded in a finer matrix.

Contact Fault: A fault complex well defined in eastern and northern Prince William Sound but seems to disappear in the western Sound. Thought to be a suture zone between the Chugach and Prince William terranes by those geologists who believe these rock groups were rafted in independently aboard the Pacific Plates.

Contact zone: A zone of baked sedimentary rock surrounding an intruded pluton. Rocks in this zone are said to be altered by contact metamorphism.

Convergent margin: A continental margin where two plates meet at a subduction zone.

Country rock: The preexisting sedimentary rock intruded by a pluton.

Dike: A magmatic intrusion that cuts across the bedding planes of the country rock. Depending upon their cooling history, dikes may exhibit chilled margins.

Exfoliation: The flaking off of large concentric sheets from a granite pluton often forming a dome-like mountain. Thought to be the result of frost:thaw action and perhaps a release of pressure from the unroofing of the pluton.

Farallon Plate : The ancestral Pacific Plate which carried the Superterrane toward the North American Continent. A final fragment of this plate, the Juan de Fuca plate, is presently subducting off the Oregon, Washington, and Northern California Coast accounting for recent volcanic activity in that region.

Fault: A shear zone in the Earth's crust where rocks have slid past each other. Faults may be regional features or limited to a small outcrop.

Feldspar: A complex aluminum silicate containing calcium, sodium and\or potassium usually formed in igneous rocks. Feldspar is usually clear, white, or gray in color. One of the most common minerals constituting continental, felsic crust.

Flute cast: Bulbous protrusions flattened on one end occurring on the bottom sandstone layer of a graded turbidite bed. Flute casts are thought to be due to the scouring of turbidity currents of the soft, muddy, upper layer of the previous turbid flow followed by infilling of the scour marks.

Flysch: Another term for turbidite deposits.

Foraminifera: Tiny, one celled microorganisms with carbonate skeletons which contribute to the formation of limestones. Foraminifera microfossils are sometimes age diagnostic and have been used to date some of the rocks of Prince William Sound.

Fractionate: The segregation of minerals in a plutonic melt under the influence of gravity or thermal diffusion.

Gabbro: Basaltic magma that has cooled slowly beneath the Earth's crust. Gabbro resembles granite in being composed of large interlocking crystals but contains far more of the darker mafic minerals.

Gneiss: A coarse-grained, banded regionally metamorphosed rock.

Graded bedding: Bedding common to most turbidite deposits where grain size grades from coarse at the bottom of the bed to fine at the top. Typically conglomerates grade from the bottom of the bed upward to coarse sandstone, then fine sandstone and finally mudstone at the top of the bed.

Granite: An igneous intrusive rock composed of large interlocking crystals, some of which are black, giving the rock a speckled appearance. Plutons are mostly granite but often contain diorite, and granodiorite ("granitic" rocks) and occasionally gabbro. The extrusive equivalent of granite is the lava "rhyolite."

Granodiorite: Similar to granite but containing relatively more darker mafic crystals. The extrusive equivalent of granodiorite is the lava, andesite.

Gravity high: Areas above the Earth's surface where gravity measurements are anonymously elevated because of the density of the underlying rocks. Typically ocean crustal rocks yield higher readings than continental rocks.

Graywacke: A hard sandstone consisting of angular fragments of quartz and feldspar in a clayey matrix: hence a dirty sandstone. The poor sorting of sand and mud particles giving fresh samples their gray color is probably due to the turbulent mixing occurring in a turbidity current. Most of the sandstones of Prince William Sound are slightly metamorphosed graywackes.

Greenschists: A greenish, foliated metamorphic rock whose color is due to the formation of the greenish minerals epidote and chlorite under conditions of relatively low temperatures and pressure. Greenschists form a) in subduction zones, b) under conditions of regional metamorphism, and c) when extruded basalts react chemically with the minerals in sea water. These latter rocks in the field are often called "greenstones."

Greenstone: Basalt cf. greenschist above.

Groove cast: Casts formed on the bottom, sandstone layer of a turbidite bed formed by the sand infilling of tool marks carved by turbidity currents in the softer mud bed of the previous turbidity flow.

Hornfels: A fine-grained metamorphic rock composed of baked sediments found in the contact zone around plutons or intruded dikes and sills.

Igneous rocks: Rocks formed from the cooling of a magma. Igneous rocks are generally composed of crystals which form when the melted rock cools. Major igneous rocks in Prince William Sound are basalts (greenstones) and granitic rocks.

Inclusions: Here, sedimentary (country) rock that has fallen into a magma so that it appears as a thermally metamorphosed fragment in the resulting granite.

Intrusives: Molten, igneous rock which has invaded preexisting sedimentary rocks. Intrusives in Prince William Sound consist of granite plutons and associated dikes and sills and basalts intruded into sediments in greenstone areas.

Kula Plate: The pre-Pacific Plate on which the present terranes of Southern Alaska were rafted into their present positions. Although "Kula" means "all gone," because this plate was thought to have been entirely subducted; the crust of the Bering Sea Basin may be a remnant. Also, a final small fragment of this plate has been discovered recently subducting in the Aleutian trench near the Aleutian Island, Attu.

Limestone: A calcium carbonate rock usually formed from shells and skeletal remains of shallow marine organisms. In Prince William Sound small, sparse limestone deposits are thought to result from the fragmented remains of these organisms carried down into a subduction trench by turbidity currents.

Lithified: The process by which sediments are transformed into rock.

Mafic: Rocks having higher concentrations of the heavier elements magnesium and iron and lower concentrations of the lighter elements aluminum, silicon, sodium and potassium which make up "felsic rocks." Mafic rocks typical of ocean crust are typically dark and heavy and have been derived from the Earth's mantle.

Magma: Molten rock.

Mantle: The main bulk of the Earth between the crust and the core made up of mafic minerals.

Matrix: Here, the finer material in which the larger clasts of a conglomerate, breccia, or melange are embedded.

McHugh Complex: A group of rocks near Anchorage composed of a tectonic trench melange of the Border Ranges Trench. The oldest portion of the Chugach Terrane.

Melange (tectonic): A group of rocks formed in a subduction trench which exhibit a block in matrix structure. The blocks may consist of local continental material deposited in the trench or rocks rafted in on the ocean plate.

Metamorphic rock: Rock which has been recrystallized by heat and/ or pressure while still in a solid state. Igneous, sedimentary or metamorphic rock can all be metamorphosed. Contact metamorphism results from heating of the country rock by the cooling of intruded magmas. Regional metamorphism results from forces at work during accretion, subduction and uplift. Metamorphic rocks often exhibit an alignment of crystals, cleavage and sometimes foliation or banding.

Mid-Ocean Ridge: An area where ocean plates are rifting apart allowing hot, basaltic magma to extrude onto the ocean floor. The ridge forms as a chain of volcanic mountains along the rifting zone. Cf. "Spreading center."

Mudstones: Fine-grained, sedimentary rocks formed from the lithification of muddy sediments.

Obduction: The transference of crustal fragments from an ocean plate to a continental plate across a subduction zone.

Ocean trench: A subduction zone where plates collide. Typically, the plate composed of the denser crust dives beneath the less dense plate to be consumed deep in the mantle. If two ocean plates collide, an Island arc is formed. Where an ocean plate dives beneath a continental plate, a magmatic arc forms inland. Ocean trenches are areas of intense seismic activity.

Ophiolite: A rock formation which seems to be a sliver of ocean crust emplaced on land. A complete ophiolite sequence consists of an upper layer of chert formed from the lithification of deep ocean sediments, followed by a layer of pillow basalts which are themselves underlain by a sheeted dike sequence which in turn is underlain by layered gabbros.

Orca group: Rocks in Prince William Sound typified by the rocks around Orca Inlet. Rocks south and west of the Contact Fault. The Orca group rocks are thought to be of Paleocene to Eocene in age as opposed to the Valdez group which are thought to be Cretaceous.

Orocline (Alaskan): The Bend in the continental margin in the region of Prince William Sound.

Paleomagnetism: Tiny magnetic mineral grains in certain rocks act like compass needles indicating the direction of the magnetic pole at the time of formation. Scientists use this information to deduce the latitude at which a rock group was formed and the amount of rotation it has undergone.

Pangaea (All earth): The ancient supercontinent which formed between 400 and 300 million years ago and began to break up about 200 million years ago. Our present continents are thought to be fragments of Pangaea.

Panthalassa (All oceans): The single, all encompassing ocean that surrounded Pangaea.

Pegmatite: A very, coarse-grained intrusive, igneous rock often found near plutons. Pegmatite dikes often contain large collectible crystals.

Peridotite: A dark, heavy, coarse-grained, ultramafic, intrusive rock thought to be a major constituent of the mantle.

Pillow basalts: Extruded submarine basalts assuming a distinctive pillow shape. Sometimes referred to as "ellipsoidal lavas" or "greenstones."

Plates: Rigid slabs of lithosphere on which the continents and oceans ride that slide over the mobile asthenosphere causing continental drift.

Pluton: A large, intrusion of igneous rock, which cools slowly beneath the Earth's surface.

Prince William Terrane: The Orca group of rocks in Prince William Sound plus the Ghost Rocks and Sitkalidak formations on Kodiak Island. Some geologists consider these rocks to constitute an independent terrane while others consider them to be a younger part of the Chugach Terrane.

Quartz: Crystalline silica (SiO_2). A hard, usually white rock occurring as intrusive veins in other rocks. One of the most common minerals of continental crust.

Reverse fault: A fault in which the block above the fault plane is thrust over the block lying under the fault plane.

Rip-up: Lithified mud fragments appearing as inclusions in the lower, graded bed of a turbidite sequence. Thought to be the result of turbulent scouring of a pliable mud bed left by preceding turbidity flow.

Sandstone: A sedimentary rock of sand-sized grains (visible to the naked eye) cemented together by a finer matrix. Most sandstones in Prince William Sound are weakly metamorphosed, have a silty matrix, and are known as graywackes. By studying the grains in a sandstone, geologists can make deductions concerning the source rocks, the transport history of the sediment, and the conditions under which a sandstone formed.

Schist: A foliated, metamorphic rock.

Sedimentary prism: Cf. accretionary prism

Sedimentary rock: Rock formed the cemented fragments of other rocks. Usually formed in a subaqueous environment.

Shale: A common unmetamorphosed mudstone (siltstone or claystone) usually exhibiting partings between bedding planes.

Sheeted dikes: Repeated sheetlike intrusions of basalt underlying the pillow basalt layer of an ophiolite sequence thought to represent the feeder vents supplying the extruded pillows.

Silica: Silicon dioxide (SiO_2). Most common sands and the mineral quartz are composed of silica.

Siliceous: High in silica.

Sill: An intrusion parallel to the bedding planes of the intruded country rock.

Slate: Metamorphosed shale with a cleavage usually oblique to the bedding planes. The most common mudstone in Prince William Sound.

Sole markings: Scour or tool marks left by turbidity currents cf. flute casts, groove casts.

Spreading Center: A rifting zone where the Earth's crust is stretched and thinned allowing the extrusion of basaltic lavas.

Strike-slip fault: A fault zone where crustal blocks slide horizontally past one another. Also called a transform fault.

Subduction zone: A region marked by an ocean trench where plates are colliding. Typically, the denser plate dives beneath, or is "subducted" beneath the less dense plate.

Superterrane: The amalgamated Alexander, Peninsular and Wrangellia terranes which now form most of southern Alaska.

Telescoping: The tendency of a terrane strike-slipping along a transform boundary to become elongated through interterrane faulting.

Terrane: A group of rocks of regional extent bounded by faults which differs markedly from those groups on opposite sides of those fault boundaries. Here, often used in the sense of "exotic terranes" far traveled crustal blocks which when emplaced exhibit the above characteristics.

Thrust fault: A fault in which the upper block is thrust over the lower block at a relatively low angle. Characteristic of faults occurring at convergent boundaries and in accretionary wedges.

Transform faults : cf. strike-slip faults.

Turbidites: Repetitive sequences of alternating graded beds of uneven thickness of slates, graywackes and sometimes conglomerates formed by the action of turbidity currents.

Turbidity currents: Submarine, sediment-laden density currents which transport great quantities of sediments down the continental slope and onto the deep ocean floor depositing them in great fan-shaped piles.

Ultramafic: Rocks rich in heavier minerals iron and magnesium. Ultramafic rocks appear to be formed from minerals originating in the Earth's mantle.

Valdez group: A group of Cretaceous rocks typified by those around the town of Valdez. Most of the Chugach and Kenai mountains are classified as Valdez Group rocks. The Valdez Group is seen as being bounded by the Border Ranges Fault to the north and the Contact Fault to the South.

Vergent: A fold is is said to be "landward vergent" when its axial plane extends in a landward direction, and "seaward vergent" when its axial plane extends in a seaward direction.

Bibliography

Ben-Avraham, Z., and A. Nur. 1983. The effect of collisional processes at subduction zones on tectonic events in the oceans. p. 22-23. In Proceedings of the Circum-Pacific Terrane Conference, edited by Howell, D.G., Jones, D.L., Cox, A., and Nur, A. Stanford, California. Stanford University Press.

Bol, A.J., 1987. Paleomagnetism of pillow basalt and sheeted dikes, Resurrection Peninsula, Alaska: Confirmation of large northward displacement in S. Alaska since early Tertiary: Geological Society of America Abstracts with Programs, v. 19, p. 594.

Bruns, T., 1983. A model for the origin of the Yakutat block, an accreting terrane in the northern Gulf of Alaska: Geology, v. 11, p. 718-721.

Byrne, T., 1979. Late Paleocene demise of the Kula-Pacific spreading center: Geology. v. 7, p. 341-344.

Byrne, T., 1986. Eocene underplating along the Kodiak Shelf, Alaska: Implications and regional correlations: Tectonics. v. 5, p. 403-421.

Byrne, T. and J. Hibbard. 1987. Landward vergence in accretionary prisms: The role of the backstop and thermal history. Geology. v. 15, p. 1163-1167.

Capps, S. R., and B. L. Johnson. 1915. The Ellamar district, Alaska: U.S. Geological Survey Bulletin 605, 125 p.

Coe, R. S., Globerman, B. R., Plumley, P. W., and Thrupp, G. A., 1985. Paleomagnetic results from Alaska and their tectonic implications: American Association of Petroleum Geologists Circum-Pacific Earth Sciences Series 1, p. 85-108.

Coney P.J., and D.L. Jones. 1985. Accretion tectonics and crustal structure in Alaska: Tectonophysics, v. 119, p. 265-283.

Cooper, A.K., Scholl D.W., and M.S. Marlow. 1976. Plate tectonic model for the evolution of the Eastern Bering Sea Basin: Geological Society of America Bulletin, v. 87, p. 1119-1196.

Cowan, D.S., 1982. Geological evidence for post-40 Ma large-scale, northwestward displacement of part of southeastern Alaska: Geology. V.10, p. 309-313.

Csejtey, B., Jr., Cox D.P. Evarts, R.C., Stricker, G.D. and H.L. Foster. 1982. The Cenozoic Denali fault and accretionary development of southern Alaska: Journal of Geophysical Research. v. 87, p. 3741-3754.

Davies, D. L., and Moore, J. C., 1984. 60 m.y. intrusive rocks from the Kodiak Islands link the Peninsular, Chugach, and Prince William terranes: Geological Society of America Abstracts with Programs, v. 16, p. 277.

Davis, A.S. and G. Plafker. 1986. Eocene basalts from the Yakutat terrane: evidence for the origin of an accreting terrane, southern Alaska: Geology, v. 14, p. 963-966.

DeBari, S.M. and R.G. Coleman. 1989. Examination of the deep levels of an island arc: evidence from the Tonsina ultramafic-mafic assemblage, Tonsina, Alaska: Journal of Geophysical Research, v. 94, No. 84, 4373-4391.

Dumoulin, J. A., 1987. Sandstone composition of the Valdez and Orca Groups, Prince William Sound, Alaska: U.S. Geological Survey Bulletin 1774, 37 p.

Dzulynski S. and E.K. Walton. 1965. Sedimentary features of flysch and graywackes. Amsterdam, London and New York, Elsevier Publishing Co.

Engebretson, D., Debiche, M., and Cox, A. 1983. Plate Motions, Paleomagnetism, and Terrane Displacement Histories. pp. 83-85. In Proceedings of the Circum-Pacific Terrane Conference, edited by Howell, D.G., Jones, D.L., Cox, A., and Nur, A. Stanford, California. Stanford University Press.

Fechner S.A. and K.J. Krause. 1987. Mineral occurrences in Port Valdez, Alaska: Geological Studies of Critical Areas, State of Alaska DNR Division of Geology and Geophysics — Public Data File 87-29.

Gardner, M.C., Bergman, S.C., Cushing, G.W , Mackevett, Jr. E. M., Plafker, G., Campbell, R.B., Dodds, C J., McClelland, W.C.,and P.A. Mueller. 1988. Pennsylvanian pluton stitching of Wrangellia and the Alexander terrane, Wrangell Mountains, Alaska: Geology v. 16, p. 967-971.

Gergen, L.D. and G. Plafker. 1988. Petrography of sandstones of the Orca group from the southern trans-Alaskan crustal transect (TACT) Route and Montague Island: Geologic studies in Alaska by the U.S. Geological Survey during 1987: U.S Geological Survey Circular 1016, p. 156-159.

Gibbons, H., 1988. Microstructures and metamorphism in an accretionary prism in Prince William Sound, Alaska. MA Thesis. UC, Santa Cruz. 191 p.

Gordon and Jurdy. 1986. Cenozoic Global Plate Motions: Journal of Geophysical Research. v. 91, No. 12, p. 395-406.

Grant, U.S., and D. F. Higgins. 1913. Coastal glaciers of Prince William Sound and Kenai Peninsula: USGS Bulletin 526.

Grant, U. S., and D. F. Higgins. 1910. Reconnaissance of the geology and mineral resources of Prince William Sound, Alaska: U.S. Geological Survey Bulletin 443, 89 p.

Gromme, S., and J.W. Hillhouse. 1981. Paleomagnetic evidence for northward movement of the Chugach terrane, southern and southeastern Alaska. In Albert, N.R.D., and Hudson, Travis, eds., The United States Geological Survey in Alaska — Accomplishments during 1979: U.S. Geological Survey Circular 823-B, p. B70-B7

Harbert, W., 1987. New Paleomagnetic data from the Aleutian Islands: implications for terrane migration and deposition of the Zodiac fan. Tectonics 6 (5):585-602.

Helwig, J. and P. Emmet. 1981. Structure of the early Tertiary Orca Group in Prince William Sound and some implications for the plate tectonic history of southern Alaska: Journal of the Alaska Geological Society, v. I, p. 12-35.

Hillhouse, J. W., Gromme, C. S., and Csejtey, B., Jr., 1985. Tectonic implications of paleomagnetic poles from lower Tertiary volcanic rocks, south central Alaska: Journal of Geophysical Research. v. 90, No. 12, p. 523-535.

Hoekzema, R.B., Fechner, S.A., and Kurtak J.M. 1987. Evaluation of selected lode gold deposits in the Chugach National Forest, Alaska: Bureau of Mines Report of Investigations/1987. IC:9113. 62 p.

Hollister, L.S. 1979. Metamorphism and crustal displacements — new insights: Episodes, v. p. 3-8.

Hudson, T., Plafker, G., and Peterman, Z. E., 1979. Paleogene anatexis along the Gulf of Alaska margin: Geology v. 7, p 573-577.

Jacob, K. H., Nakamura, K. and Davies, J. N., 1977. Trench-volcano gap along the Alaska-Aleutian Arc — Facts and speculations on the role of terrigenous sediments for subduction. In Talwani M., and Pitman, in. C., III eds., Island arcs and deep sea trenches and back arc basins: American Geophysical Union, Maurice Ewing Series. 1:243-258.

James, S. A., Hollister L.S., and Morgan W.J., 1989. Thermal modeling of the Chugach metamorphic complex: Journal of Geophysical Research, v. 94, No. 84, p. 4411-4423.

Jansons, Uldis, Hoekzema, R. B., Kurtak, J M., and Fechner, S. A., 1984. Mineral occurrences in the

Chugach National Forest, Southcentral Alaska: U.S. Bureau of Mines Open-File Report MLA 5-84,

Johnson, B. L., 1914. The Port Wells gold-lode district: U.S. Geol. Survey Bull. 592, p. 195-236.

___ 1916. The gold and copper deposits of the Port Valdez district: U.S. Geological Survey Bulletin 622, p. 140-188.

___ 1918. Copper deposits of the Latouche and Knight Island districts, Prince William Sound: U.S. Geological Survey Bulletin 662, p. 193-220.

Jones, D.L. , Blake, M. C., Jr., Bailey, E. H., and R.J. Mclaughlin. 1978. Distribution and character of upper Mesozoic subduction complexes along the west coast of North America: Tectonophysics, v. 47, p. 207-222.

Jones, D.L., Cox, A., Coney, P. and Beck, M., 1982. The Growth of Western North America: Scientific American, Vol. 247, No. 5, p. 70-84.

Jones, D. L., N. J. Silberling, and J. W. Hillhouse. 1977. Wrangellia — A displaced terrane in northwestern North America: Canadian Journal of Earth Sciences, v. 14, p. 2565-2577.

Jones, D.L., Silberling, N.J., Berg, N.C., and G. Plafker. 1981. Map showing tectonostratigraphic terranes of Alaska, columnar sections, and summary description of terranes: U.S. Geological Survey Open-File Report 81-792, 20 p., 2 pls., scale 1:2,500,000.

Keller, Gerta, von Huene, Roland, McDougall, Kristin, and T.R. Bruns. 1984. Paleoclimatic evidence for Cenozoic migration of Alaskan terranes: Tectonics. 3:473-495.

Kurtak, J.M. and R.E. Jeske. 1986. Mineral Investigations in the Chugach National Forest, Alaska (islands area). DOI, Bureau of Mines: OFR 54-86. 302 p.

Lethcoe N.R. 1987. An Observer's Guide to the Glaciers of Prince William Sound. Prince William Sound Books. Valdez, Alaska.

Little, T.A., 1988, Tertiary tectonics of the Border Ranges fault system, north central Chugach Mountains Alaska: Sedimentation, deformation and uplift along the inboard edge of a subduction complex, Ph. D. thesis, 343 pp. Stanford University., Stanford, Calif.

Lonsdale, P. 1988. Paleogene history of the Kula plate: Offshore evidence and onshore implications: Geological Society of American Bulletin. v. 5, p. 733-754.

McGlasson, J. A., 1976. Geology of central Knight Island, Prince William Sound, Alaska: Colorado School of Mines, Golden, Colorado, M. S. thesis, 136 p.

Molnia, B.F. 1986. Glacial history of the northeastern Gulf of Alaska — a synthesis. In Glaciation in Alaska. eds. T.D. Hamilton, Reed, and Thorson. Alaska Geological Society.

Moffit, F. H., 1954. Geology of the Prince William Sound region, Alaska: U.S. Geological Survey Bulletin 989-E, p. 225-310.

Moffit, F. H., and R. E. Fellows. 1950. Copper deposits of the Prince William Sound district, Alaska: U.S. Geological Survey Bulletin 963-B, p. 47-80.

Moore, J.C., Byrne, T., Plumley, P.W., Reid, M., Gibbons, H., and Coe, R.S., 1983. Paleogene evolution of the Kodiak Islands, Alaska — consequences of ridge-trench interaction in a more southerly latitude: Tectonics, v. 2, p. 265-293.

Nelson, S.W. 1987. The Midas Mine: A stratiform Fe-Cu-Zn-Pb sulfide deposit in late Cretaceous turbidite near Valdez, Alaska. 1987 GSA Abstract.

Nelson, S.W., Blome, C.D., Harris, A.H., Reed, K.M., and Wilson, F.H., 1986. Late Paleozoic and Early Jurassic fossil ages from the McHugh Complex. In Bartsch-Winkler, Susan, and Reed, K.M., eds., Geologic Studies in Alaska by the U.S. Geological Survey during 1985: U.S. Geological Circular 978, p. 60-64.

Nelson, S. W., Dumoulin, J. A., and Miller, M. L., 1985. Geologic map of the Chugach National

Forest, Alaska: U.S. Geological Survey Miscellaneous Field Studies Map MF-1645-B, 16 p., pl., scale 1:250,000.

Nelson, S.W., and Hamilton, T.D., 1989. Guide to the geology of the Resurrection Bay-eastern Kenai Fjords area. Alaska Geological Society. Anchorage, Alaska.

Nelson, S.W., Miller, M.L., Barnes, D.F., Dumoulin, J.A., Goldfarb, R.J., Koski, R.A., Mull, C.G., Pickthorn, W.J., Jansons, Uldis, Hoekzema, R.B., Kurtak J.M., and Fechner, S.A., 1984. Mineral resource potential of the Chugach National Forest, Alaska: U. S. Geological Survey Miscellaneous Field Studies Map MF-1645-A, 24 p., 1 pl., scale 1:250,000.

Newton. C. R. Paleozoogeographic affinities of Norian bivalves from the Wrangellian, Peninsular and Alexander terranes western North America . In C. H. Stevens ed. Pre-Jurassic rocks in western North American suspect terranes: Society of Economic Paleontologists and Mineralogists, Pacific section, p. 37-48.

Nokleberg, W. J., Jones, D. L., and Silberling, N. J., 1985, Origin, migration, and accretion of the Maclaren and Wrangellia terranes, eastern Alaska Range, Alaska: Geological Society of America Bulletin, v. 96, p. 1251- 1270.

Panuska, B. C., 1985. Paleomagnetic evidence for a post-Cretaceous accretion of Wrangellia: Geology, v. 13, p. 880-883.

Pavlis, T. L., 1982. Cretaceous deformation along the Border Ranges fault, southern Alaska — brittle deformation during formation of a convergent margin: Tectonics, v. 1, p. 343-368.

Plafker, G., 1969. Tectonics of the March 27, 1964, Alaska earthquake: U.S. Geological Survey Professional Paper 543-I, p. 11-174.

Plafker, G., 1988. Regional geology and petroleum potential for the northern Gulf of Alaska continental margin. In Scholl, D.W., Grantz, A., and Vedder, J.G., eds. Geology and resource potential of the continental margin of western North America and adjacent ocean basins — Beaufort Sea to Baja California: Circum-Pacific Council for Energy and Mineral Resources Earth Science Series, Houston, Texas. v. 6, p. 229-268.

Plafker, G., and F. S. MacNeil, 1966. Stratigraphic significance of Tertiary fossils from the Orca Group in the Prince William Sound region, Alaska: Geological Survey research 1966: U.S. Geol. Survey Prof. Paper 550-B, p. B62-B68.

Plafker, G., Noklelberg W. J., and J.S. Lull, 1989. Bedrock geology and tectonic evolution of the Wrangellia, Peninsular, and Chugach terranes along the trans-Alaska crustal transect in the Chugach mountains and southern Copper River Basin, Alaska: Journal of Geophysical Research, v. 94, No. 84, p. 4255-4295.

Plafker, G., Keller, G., Barton, J. A., and J. R Elueford. 1985. Paleontologic data on the age of the Orca Group, Alaska: U.S. Geological Survey Open-File Report 85- 429, 24 p.

Plafker, G., Keller, G., Nelson, S. W., Dumoulin, J. A., and Miller, M. L., 1985. Summary of data on the age of the Orca Group. In Bartsch-Winkler, S. W., ed., The United States Geological Survey in Alaska — Accomplishments during 1984: U.S. Geological Survey Circular 967, p. 74-76.

Plumley, P. W., Coe, R. S., and Byrne, T., 1983. Paleomagnetism of the Paleocene Ghost Rocks Formation, Prince William Terrane, Alaska: Tectonics, v. 2, p. 295-314.

Roeske, S.M., Mattinson, J.M., Armstrong, R.L. 1989. Isotopic ages of glaucophane schists on the Kodiak Islands, southern Alaska, and their implications for the Mesozoic tectonic history of the Border ranges fault system: Geological Society of America Bulletin. 5, p. 1021-1037.

Sawkins, F.J., Chase C.G., Darby, D.G. and Rapp G., 1978. The Evolving Earth. New York, Macmillan.

Silberman, M.L., Mackevett, Jr. E.M., Connor, C.L., Klock, P.R., Malechitz, G. 1979. K-Ar ages of the Nikolai greenstone from McCarthy quadrangle, Alaska — the "docking" of Wrangellia: US Geological Survey in Alaska, Accomplishments: 1979. B61-B63.

Stefansson, K., and R. M. Moxham. Copper Bullion Claims Rua Cove, Knight Island, Alaska: U.S. Geol. Surv. Bull. 947-E, 1946, p. 85-92.

Stone, D.B., Panuska, B.C., and Packer, D.R., 1982. Paleolatitudes versus time for southern Alaska: Journal of Geophysical Research, v. 87, p. 3697- 3707.

Tarr, R.S., and L. Martin. 1914. Alaska glacier studies of the National Geographic Society in the Yakutat Bay, Prince William Sound and Lower Copper River Regions. The National Geographic Society.

Thrupp, G. A., and R. S Coe. 1986, Early Tertiary paleomagnetic evidence and the displacement of southern Alaska: Geology, v. 14, p. 213-217.

Tysdal R. G., and J. E. Case. 1979. Geologic map of the Seward and Blying Sound quadrangles, Alaska: U.S. Geological Survey Miscellaneous Geclogic Investigation Series Map I-1150, 12p., 1 pl., scale 1:250,000.

Von Heune, R., Keller G., Bruns T.R. and McDougall K., 1985. Cenozoic migration of Alaskan terranes indicated by paleontologic study. In Howell G. ed. Howell, David G. Tectonostratigraphic Terranes of the Circum-Pacific Region: Circum-Pacific Council for energy and mineral resources, p. 121-136.

Wallace, W.K. and D.C. Engebretson. 1984. Relationships between plate motions and late Cretaceous to Paleogene magmatism in southwestern Alaska: Tectonics. 3(2):295-315.

Williams, F. E. 1961. A preliminary study of the feasibility of fracturing the Copper Bullion deposit, Alaska, with nuclear explosives: BU Mines Unpubl. Rep., available upon request from Anchorage, AK.

Winkler, G. R., 1976. Deep-sea fan deposition of the lower Tertiary Orca Group, eastern Prince William Sound, Alaska. In Miller, T. P., ed., Recent and ancient sedimentary environments in Alaska: Anchorage, Alaska Geological Society Symposium Proceedings, p. R1-R20.

Winkler, G. R., McLean, H., and Plafker, G., 1976, Textural and mineralogical study of sandstones from the onshore Gulf of Alaska Tertiary Province, southern Alaska: U.S. Geological Survey Open-File Report 76- 198, 51 p.

Winkler, G. R., and Plafker, G., 1981, Geologic map and cross sections of the Cordova and Middleton Island quadrangles, southern Alaska: U.S. Geological Survey Open-File Report 81-1164, 24 p.

Winkler, G. R., and G. Plafker. 1975. The Landlocked fault: Part of a major early Tertiary plate boundary in southern Alaska. In Yount, M. E., ed., The United States Geological Survey Alaska Program: U.S. Geol. Survey Circ. 722, p. 49.

Index

Accretion (terrane), 29, 30-31

Accretionary prism (wedge), 27-28, 54, 58, 61, 62

Alaska Homestake Mine, 140-141

Alaskan orocline, 36, 43, 44, 45, 60, 72

Aleutian Trench, 20, 21-23, 25, 43-45, 70, 72, 73, 74, 80, 85

Alexander Terrane, 32, 37, 38, 48, 68, 82

Bainbridge Island, 203

Bainbridge Fault, 199

Basalt, 87-91

Bay of Isles, 197

Bettles Bay, 137, 142

Bidarka Point, 177

Billings Glacier Pluton, 96, 132-133

Black smokers, 16, 17

Bligh Island, 177

Blue schists, 28, 39, 57-58

Border Ranges Trench (Fault), 26, 35-36, 39-40, 57-58, 59, 61, 66, 72, 82ff.

Boulder Bay, 177

Canoe Passage, 71

Cascade Bay, 150

Cedar Bay (Hawkins Island), 71

Cedar Bay Pluton, 65, 152, 153

Chamberlain Bay, 155

Chenega Island, 22, 204-206

Chugach Mountains, 20, 25, 76

Chugach-Prince William Terrane, 26, 32, 34, 35, 38-42, 43, 44, 58ff, 82ff, 119, 169

Cliff Mine, 75-76, 128, 139, 162-163, 173

Coghill Point, 141

College Fiord, 22

Comfort Cove, 182

Concretions, 115

Conglomerate, 10, 58, 90, 103, 104, 107-110, 131, 136, 140, 170-171

Contact Fault, 40, 41, 82, 119, 149, 155, 166

Contact Zone, 94, 100

Continental crust, 6-12

Copper, 17, 63, 87, 126, 127, 128, 163-166, 173, 177-180, 181, 194, 195-196, 197, 200-202

Cordova, 117, 183-184

Cow Pens Anchorage, 150-151

Crafton Island, 145, 206

Culross Island, 145-147

Culross Island Pluton, 73, 146-147

Culross Mine, 145-146

Dangerous Passage, 205-206

Deep Water Bay, 2, 207

Double Bay, 187

Drier Bay, 195-196

Duke and Duchess Mines, 200

Earthquake, 18-19, 22, 124, 158-159, 190-191

Ellamar Mine, 17, 87, 173

Elrington Island, 202-203

Eshamy Pluton, 73, 97

Esther Island Pluton, 73, 142-145

Evans Island, 90, 111, 203

Faults, 7, 123-126

Feldspar, 7, 135

Felsic Rocks, 7, 13, 92, 93

Fidalgo Mining Co. Mine, 181

Flute casts, 111-112

Flysch, 11, 26, 54, 102

Folds, 7, 26, 121-123

Fossils, 82, 84, 117-119, 135, 136, 141, 168, 169

Gabbro, 8, 93, 97, 142
Galena Bay, 12, 78, 90, 107, 110, 118,
    155, 168-173
Garden Cove, 187
Gibbon Anchorage, 188-189
Gold, 23, 26, 75-76, 128, 129, 133, 134,
    137-138, 138-139, 140-141, 142, 145-
    146, 156, 160-162, 173
Golden, 136, 142
Gold King Mine, 156
Graded bedding, 10, 103-107
Granite, 7, 23, 136
Granite Bay (Dangerous Passage), 206
Granite Mine, 136, 138-139
Gravina-Nietzatin Terrane, 32, 38, 54-55,
    57
Graywacke, 9, 58, 90, 103, 133
Green Island, 105, 108, 110, 116, 188-189
Greenschists, 281
Greystone Bay, 97
Groove Casts, 113
Hanning Bay, 190
Harriman Fiord, 135, 136, 140-141
Harrison Lagoon, 136, 139
Hawkins Island, 119, 183
Hinchinbrook Island, 118, 123, 185-187
Hobo Bay, 138-139, 142
Hogg Bay, 203
Horseshoe Bay, 200
Igneous rock, 8, 86
Intrusives, 8, 86, 90, 92-97
Irish Cove, 181
Fish Bay, 180
Growler Bay, 155
Jack Bay, 75, 166
Jack Bay Fault, 166
Johnstone Point, 187
Knight Island Group, 17, 63-64, 79 89,
    90, 91, 93, 123, 192-197
Kula Plate, 30, 59ff., 85ff
Landlocked Bay Mines, 126, 128, 177-
    180

Lansing Mine, 134
Latouche Island, 199-202
Latouche Mines (Beatson, Chenega, Gird-
    wood), 17, 200-202
Limestone, 7, 48, 50, 87
Load casts, 115
Lone Island, 147
Long Bay (near Columbia Glacier), 62,
    155
Lower Passage, 195

Mafic rocks 8, 13, 92, 93
Main Bay
McHugh Complex, 26, 39, 58, 82
Melange, 26-27
Metamorphic rock, 7, 23, 86, 94, 97-101
Midas Mine, 17, 163-166, 173
Mid-ocean spreading center (ridge), 15,
    16, 18, 19, 49, 62ff, 89-90
Mineral Creek Mining Area, 160-162
Mineral King Mine, 137-138
Montague Island, 123, 126, 187-190

Naked, Peak & Storey Islands, 78, 109,
    118, 152, 153-154

Ocean crust, 9-13, 16-18, 19, 23
Ophiolite, 12, 13, 16, 63, 91, 193-194
Orca Group, 41, 42, 60, 62, 82, 102, 119,
    155, 167, 169
Outpost Island, 36

Paleomagnetism, 31, 34-35, 39, 72
Pandora Group Mines, 197
Peninsular Terrane, 32, 35, 37, 38, 39, 50-
    53, 54, 58, 65, 82ff.
Peridotite, 6, 7
Passage Canal, 131-134
Passage Canal-Port Wells Pluton, 73, 74,
    97, 133-134
Perry Island, 147
Pigot Bay, 134-135

Pillow Basalts, 12, 62, 87-89, 170, 193-194
Plutonic rock, 8, 86, 92-97
Plutons, 8, 23, 28, 51, 64, 65, 73, 92-97
Poe Bay, 131
Point Doran, 136, 140
Point Nowell, 206
Point Whiteshed, 184
Portage Bay Mine, 133-134
Port Fidalgo, 78, 177-181
Port Gravina, 78, 181-183
Port Nellie Juan, 1, 23, 24, 73, 79, 92, 207
Port Wells, 134, 136-142

Quartz, 7, 23, 131, 133, 137

Ramsay Rutherford Mine, 160
Rip-up, 110
Resurrection Peninsula Ophiolite, 63-64
Rua Cove Mine, 17, 63, 194
Ruff'n Tuff Mine, 156

Sawmill Bay (Valdez Arm), 168
Schlosser Mine, 181
Sedimentary rock, 7, 910, 23, 86, 102-126
Serpentine Cove, 140
Shale, 7, 58
Sheep Bay Pluton, 65, 96-97, 181-182
Sheeted dikes, 12, 89=91-194
Shelter Bay (Hinchinbrook), 187
Simpson Bay, 109, 182-183
Slate, 7, 9, 27, 90, 133
Sleepy Bay, 199
Snug Harbor, 196
Squire Island, 196
Squaw Bay, 149
Stockdale Harbor, 189
Strike-slip or transform faulting, 15, 32,
    33-34, 36, 44, 67, 72
Subduction zone (trench), 15, 18, 19, 20,
    23-24, 26-27, 62ff, 91
Sunny Cove, 181
Serpentine, 37, 38, 53ff, 82ff

Surprise Cove, 136

Tatitlek Narrows, 176-177
Terrane, 28, 29-30, 33ff
Thrust faulting, 21, 22, 32, 33, 124
Turbidites, 10, 11, 62, 91, 102ff, 131
Turbidity current, 10, 11, 24, 26, 58, 62-
    63, 80, 90-91, 102ff

Unakwik Inlet, 22, 62, 118, 142, 150-151

Valdez, 17, 26, 82, 119, 142, 157-160
Valdez group, 41, 42, 59, 75, 82, 93, 102,
    119, 155, 167, 169
Volcanism, 15, 19, 22-23, 86, 87-91

Wells Bay, 142
Whale Bay, 203
Whittier, 132
Wrangellia, 32, 34, 35, 37, 38, 39, 47-50,
    53, 68, 82ff

Yakutat Terrane, 32, 37, 43-44, 75

Zaikof Harbor, 187
Ziegler Cove, 82, 118, 135